Eco-House Renovations

45 Green Home Conversions

Lucy D. Rosenfeld

Schiffer Publishing Ltd
4880 Lower Valley Road • Atglen, PA 19310

PHOTO CREDITS

Front cover: Main: Bercy Chen Studio BP, Architects (see page 66). Insets, top to bottom: Ted Montgomery, Groudswell Architects, (see page 156); Peter Kurt Woerner and Associates (see page 62); Obie G. Bowman, Architect (see page 206); Lake Flatow Architects (see page 160)

Back cover: Main: Webber and Studio, Inc. Architects (see page 232). Insets, top: Locus Architects (see page 176); Lake Flatow Architects (see page 160); bottom: Angela Dean AIA, AMD Architecture (see page 166)

Spine: Obie G. Bowman, Architect (see page 206)

Endsheets: Angela Dean, AIA, AMD Architects (see page 92)

Title page: Bercy Chen Studio LP (see page 172)

Page 8: Lake Flatow Architects (see page 160)

Part I opener: Leger Wanaselja Architects (see page 216)

Pages 238-239: Leger Wanaselja Architects (see page 216)

Page 240: Charles W. Moore and Mark Simon, Centerbrook Architects; Centerbrook Architects and Planners (see page 188)

Library of Congress Control Number: 2011942465

Designed by John P. Cheek
Cover design by Bruce Waters

Type set in Planet Estyle/Myriad Pro

ISBN: 978-0-7643-3929-5
Printed in China

Schiffer Books are available at special discounts for bulk purchases for sales promotions or premiums. Special editions, including personalized covers, corporate imprints, and excerpts can be created in large quantities for special needs. For more information contact the publisher:

Published by Schiffer Publishing Ltd.
4880 Lower Valley Road
Atglen, PA 19310
Phone: (610) 593-1777; Fax: (610) 593-2002
E-mail: Info@schifferbooks.com

For the largest selection of fine reference books on this and related subjects, please visit our website at

www.schifferbooks.com

We are always looking for people to write books on new and related subjects. If you have an idea for a book, please contact us at proposals@schifferbooks.com

This book may be purchased from the publisher.
Include $5.00 for shipping.
Please try your bookstore first.
You may write for a free catalog.

In Europe, Schiffer books are distributed by
Bushwood Books
6 Marksbury Ave.
Kew Gardens
Surrey TW9 4JF England
Phone: 44 (0) 20 8392 8585; Fax: 44 (0) 20 8392 9876
E-mail: info@bushwoodbooks.co.uk
Website: www.bushwoodbooks.co.uk

Eco-House Renovations

45 Green Home Conversions

Lucy D. Rosenfeld

Contents

A remodeled 1971 California house by Obie Bowman has newly braced frame and large windows, bringing sunlight deep into the interior. (See Chapter 39.) Photo by Tom Rider.

Introduction

WHY ECO-HOUSE RENOVATION?

Welcome to the planning of an eco-house conversion. Whether you live in a town house, a village colonial, a country barn, or a coastal saltbox, you can convert your old house into an ecologically sound new house—with the inspiration provided by the many architects included in these pages.

This is basically a book about design, rather than technology. But it will introduce you to the world of eco-conversion and the major issues such a renovation will deal with. (While there are dozens of small details and devices that can be helpful in converting a building to a greener version of itself, this book is not about little steps like changing light bulbs or setting out rain barrels, as helpful as such steps are.)

The major issues discussed in the following pages concern the basic design and how it can effectively change our world—one house at a time—from the size of the building's footprint, to its energy needs, its use of sustainable or even reused materials, and the amount of pollution emitted—all within an aesthetically pleasing and practical renovation.

The basic issues: energy efficiency, the use of nature and natural sources, conservation, recycling, clean air—these are all issues basic to the green conversion. This book is divided into two parts: in the first section we look at possible elements in a green renovation. What can be changed? How can the designer reconfigure a traditional building so it is ecologically sound? How can the newer—even recently built—structure be updated? All of the issues listed in Part I may be part of the basic planning of the renovation, for they are central to its success as an eco-conversion.

AMONG THESE BASIC CHANGES ARE:

- Finding alternative energy sources for heat, cooling, electricity, and hot water
- Using space intelligently: reconsidering the size of the building's footprint
- Planning an ecologically sound design, such as unusual roofs or solar hallways
- Reusing and salvaging of the original building, and materials for reconstruction
- Dealing with the climate of the area
- Using nature itself; from earth berms to trees to roof gardens
- Avoiding toxic chemicals in wood, paint, finishes, etc.
- Understanding glass and its surprising thermal properties
- Choosing building materials: new, old, manufactured, replenishable, sustainable
- Finding new types of insulation to create a tight "thermal envelope"
- Considering ventilation; choosing a design that brings fresh, clean air into the house
- Saving water

Of course, there are negative issues, as well. Like any new inventions and ideas, there are unexpected repercussions. For example, the new extremely tight thermal envelopes that many architects propose may leave the house without enough fresh air intake. The roof solar panels that are so successful at reducing energy needs may not be sustainable in a very snowy climate. These and other downsides to eco-design may be remedied by new technology and should be carefully considered before construction.

In Part II of the book we visit 45 projects designed by architects—homes reconfigured, renovated, added to, or otherwise redone with many of these principles in mind. They are situated all over the country, in a variety of climates, and their renovations range from simple and basic (salvaged kitchen fixtures, non-polluting finishes, rain chains) to elaborate and stunning (the solar sink, the air foil roof, the recycled power station). These are among the most interesting of new projects—both in their use of new technologies and materials and in their design elements; all of the architects were aiming for a workable combination of the aesthetically pleasing and the ecologically sound.

At the end of the book you'll find a list of organizations and seals of approval to help in a green renovation, as well as contact information about each of the architects whose work is featured.

The architects and the author hope that the ideas in this book will bring you—the renovator—to a new appreciation of the innovations and possibilities of a greener, more sustainable home.

PART I
Choices for Going GREEN

CAM

Chapter 1

Planning Your Renovation

The planners of an eco-renovation—no matter how small or how grand—should have green concepts in mind from the start. Begin by choosing an architect who thinks "green," and who will be innovative in working with you toward a sustainable and ecologically sound renovation. (Don't forget that the organizations listed at the end of the book also can help in the idea stage.)

In the following pages you'll find summaries of the most common and basic elements of eco-renovation to aid in your planning.

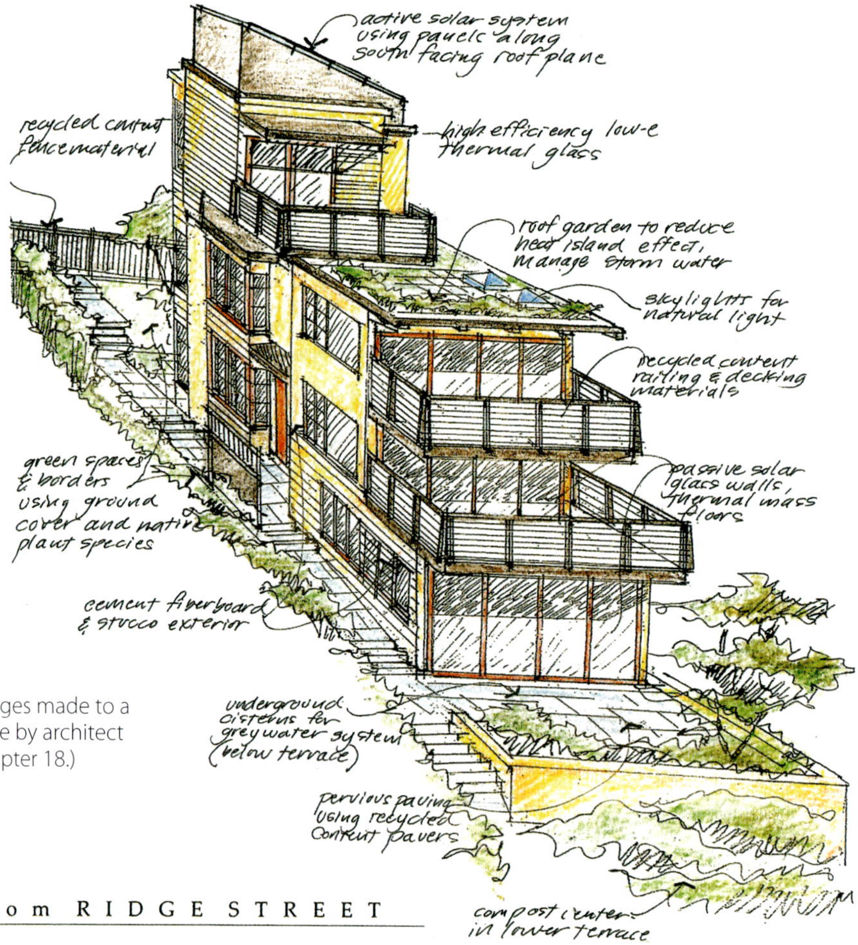

A diagram showing changes made to a Hudson River Town House by architect Christina Griffin. (See Chapter 18.)

HERE ARE SOME QUESTIONS YOU SHOULD CONSIDER DISCUSSING:

- What is salvageable? Reclaiming buildings, as well as construction material, is the most basic of ecological ideas.
- What are the costs of each change you want to make? Some eco-changes are remarkably reasonable, while others will take some years to pay for themselves. (Keep in mind that there are a variety of tax credits and governmental subsidies available for green projects.)
- In what way is your location appropriate for eco-changes? Is it warm and sunny for most of the year and ideal for solar panels? How can you create a tight thermal envelope for a cold and snowy climate?
- What local materials are easily available and will not require long truck routes, or rare, non-sustainable woods, for example?
- How will you provide energy for your house? Are sun, wind, ground pumps, or other new sources of energy efficiency feasible?
- Is the layout you have in mind as energy-efficient and is its size (footprint) as minimal as it can be?
- Have you investigated the new cutting-edge technology available to eco-builders?
- Is water conservation an issue? What simple (rain barrels) or more high-tech (water recycling systems) can you use?
- What insulation, paints, finishes, and other materials can you choose that are non-toxic and non-polluting?
- How can natural vegetation help in your plans?

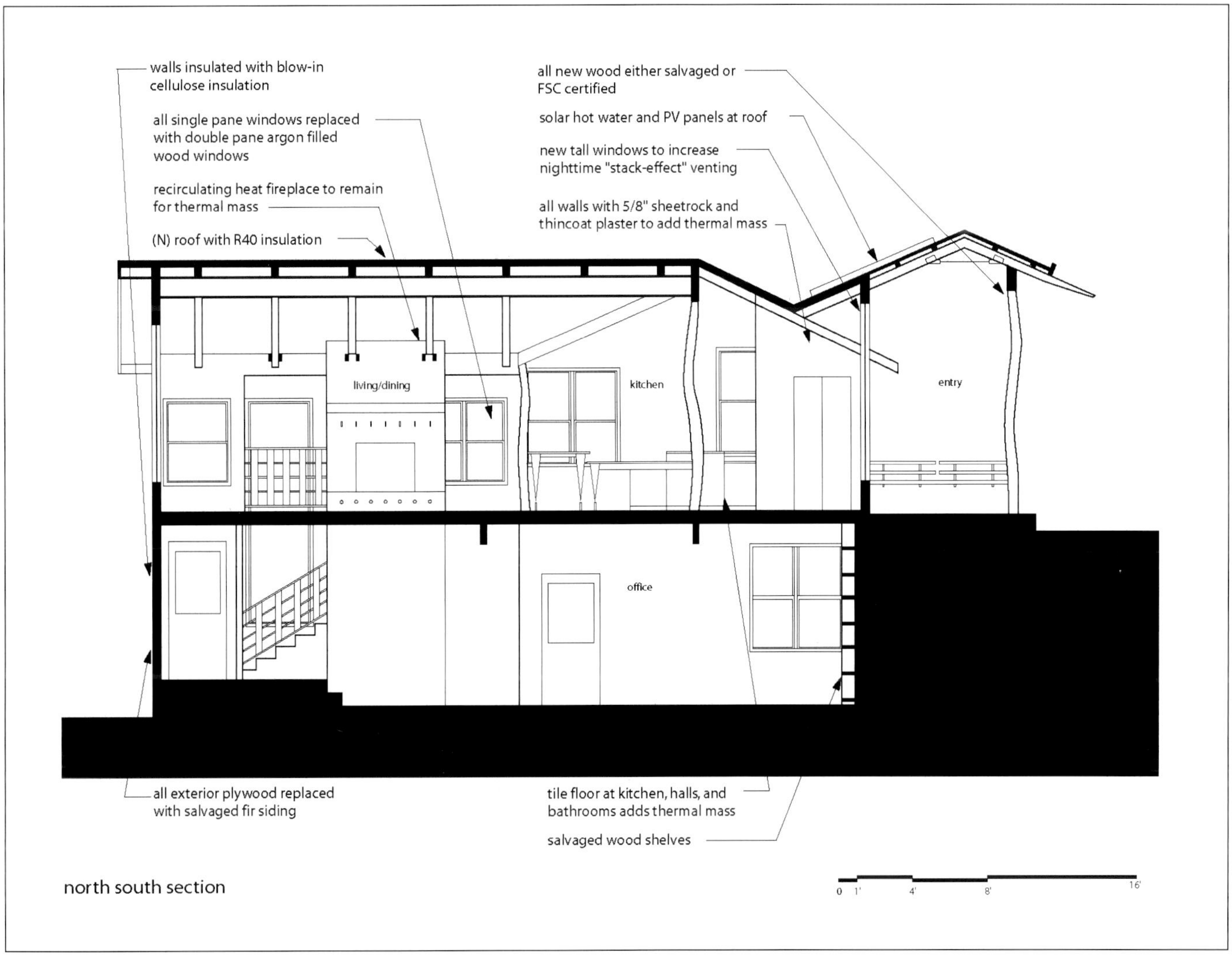

This diagram of house renovation by architects Leger Wanaselja identifies new, ecologically sound alterations. (See Chapter 24.)

Chapter 2

Recycling, Salvaging, and Renewing

Certainly one of the essential elements of the eco movement is the reuse of materials. Home renovators, as well as builders of new houses, can make tremendous gains by reusing everything from old doors to salvaged shingles, discarded windows to carefully removed roofing materials. Notable architects have made a conscious effort to salvage and recycle such elements into new uses. In fact, it is now possible in many cities to get recycled materials at a warehouse, where builder and demolition crews have left discarded materials. (Check the Building Materials Reuse Association –BMRA—which has a listing of such warehouses.)

Salvaging old materials is good for our communities in many ways: it saves landfill space, it saves the cost of new materials and their transportation to the building site, and it reduces pollution caused by many materials that do not naturally disintegrate.

Adaptive reuse—the term for reusing an old building in a new way (such as converting a spacious warehouse into living/working lofts, or a stately bank building into a downtown home) is now a fairly common phenomenon. Conversion of commercial buildings that only a few years ago might have been demolished, has become more common in the downtowns of cities. But houses, too, can be converted—from one story to two, from small to larger, from one style to another—and from energy-users to sustainable.

Salvage of building elements not only meets the aims of eco-sensitive architects, builders and home renovators; it also offers aesthetic pleasures—particularly to those who enjoy the antique or unusual. Items discarded as trash can be mixed and matched in imaginative ways; recycled items are often reused in entirely unexpected ways. All kinds of recycled or salvaged items are usable and ecologically friendly.

Most of the examples of home renovations in this book include some form of salvage and recycling. Among them are kitchen countertops made from a recycled bowling alley, a home and garden built with large rectangular shipping containers, a fireplace in an elegant Connecticut home made from salvaged tin from the shed's roof, and a variety of industrial buildings converted into homes.

Inventive Reuse: One of the most unusual reuse projects is this one by Jennifer Siegal of the Office of Mobile Design. Here, recycled, boxy cargo containers are combined with shiny new glass walls. The reworking of these large, rectangular containers (and old grain trailers that function as koi ponds in the garden) illustrate how imaginative recycling can blend in with contemporary design. Photo by Benny Chen.

Recycling Materials: A renovated barn by Centerbrook Architects features a tin fireplace made from a recycled shed roof; the original siding was salvaged for the converted barn (See Chapter 33.) Photo by Norman McGrath.

Adaptive Reuse: Reworked San Antonio industrial building was saved from demolition; architects Lake Flato retained the original steel support system and opened the interior to light with saw-tooth window arrangement. (See Chapter 26.) Photo by Paul Hester, Hester and Hardaway Photography

Salvaging Materials: Saving and reusing materials is key in this renovation by Arkin Tilt Architects. Here the kitchen counter is made from a recycled bowling alley and the sink comes from an old building. (See Chapter 12.) Photo by Edward Caldwell

Chapter 3

Renovation Footprints

Any renovation begins with the overall design. Keeping in mind the emphasis of eco-designers on smaller buildings, one of the first questions will be: can the conversion be completed within the current building perimeters, or is an addition necessary? Similar questions will include basic issues of design. Can the new rooms or technological systems be accommodated by building up or down, rather than out? Can the renovation cover a smaller area—or footprint—than the original structure? If an addition is the answer, how can the footprint be small enough to be energy efficient? What about imaginative uses for garden spaces?

Many of the examples that appear in this book have dealt with these primary issues with innovative solutions, including building on the roof, surrounding the original building with a "new" one, converting unused basement space, building interior gardens, or creating additions that add to the overall eco-friendly design of the structure.

Opposite page:
Building Up: One way to reduce the building's footprint is demonstrated by two different projects: architect Andre Tesoro's New York townhouse renovation (opposite, top) and Rick Harlan Schneider's striking house pavilion in Washington (opposite, bottom). The new additions are on the roof. (See Chapter 31and Chapter 37.) Bottom photo by Dan Redmond.

Building Down: Another way to add space without enlarging the footprint of the house is to utilize space below the house. In Obie Bowman's California renovation, a new recreation area is added below the existing house. (See Chapter 39.) Photo by Tom Rider

Building Within: In reworking an old warehouse, architect Ann Temple of SMNGA created an interior garden in the middle of the original building. (See Chapter 13.)

Chapter 4

Foundations and Floors

Basic, of course, to the thermal envelope and the ecological building system are the foundation and the flooring. Although most renovations will not involve completely redoing the foundation, there are examples of the conversion of original foundations, (even while the house is left standing), as well as the choice of new foundations for additions.

The rebuilding of foundations usually involve cement slabs with reinforcing steel or stone or wood. A frost wall (removed by several inches from the edges of the foundations) is often recommended in cold climates. The cement slab provides an insulated surface. Today's concrete is a mixture of stone and sand that hardens (and continues to get harder as time passes). The setting, thickness, and reinforcement of the slab should be related to the terrain, the climate, and the size of the footprint of the house, among other considerations. The excellent thermal qualities of concrete make it a popular eco-material, appropriate for in-floor radiant heating systems. (With wire-mesh attached directly to a subsurface of wood, or to the concrete itself, the floor surface can be installed on top of it with significant energy savings.)

Concrete is now frequently used for flooring as well as foundations. It is durable and can serve both as a foundation and as a floor surface, and it can easily be adapted for an in-floor radiant heating system. One new material favored by eco-designers is Syndecrete. It is about 40% recycled content, and is twice as light as normal concrete—although it has twice its compressive strength. It acts as a kind of "solar sink" storing up the sun's warmth and slowly releasing it at night. (For this kind of solar activity, of course, it has to be used where sunlight reaches it—such as the indoor-outdoor patio shown here.)

The use of concrete is generally thought to be ecologically sound. (It should be noted, however, that the making of concrete uses a lot of energy and spews an amount of carbon dioxide into the atmosphere, so some eco-conscious builders may combine it with one quarter of fly-ash—a bi-product of power plants.) When concrete is used for flooring, it can be mixed with tints and shined to give it the appearance of tile or stone.

Other types of eco-friendly flooring include various kinds of wood. With its natural beauty and durability, wood is still the most commonly used material for flooring. New choices of woods include the most renewable types: bamboo, ipe, teak, and "engineered woods," as well as local woods that require little transportation. Bamboo (which is actually a grass, not a tree) is becoming more and more popular. It can be harvested every three to five years (as compared with hardwoods which generally take

Foundation Before and After: A Maine renovation by architect Bob Knight reworks the uninsulated wood and concrete foundation with a new slab of concrete and indoor radiant heat. (See Chapter 22.)

Concrete Flooring: One of the newest looks in flooring is Lake Flatow's mixture of concrete and crank-case oil in this chic Texas renovation. (See Chapter 26.) Photo by Paul Hester, Hester and Hardaway Photography.

25 years to mature). Bamboo flooring comes in a variety of versions, can be glued or nailed down, and has a warm, natural tone. It can be covered with carefully chosen finishes with no toxic properties. Another choice is ipe, a wood from South America that, along with teak, is increasingly popular among eco-builders and can be obtained from LEED certified forests. What is known as "engineered wood" is an alternative to the solid hardwood floor. It has a plywood core that is covered by a hardwood veneer; it comes in tongue and groove or snap-on lengths and can be sanded or refinished.

Another option that many builders are choosing for floors is cork. Soft underfoot and able to dull loud sounds, it is now available in planks to glue or nail. Cork comes from the dead bark of oak trees and is often made from recycled materials. It comes in a variety of textures and can be dyed different tones.

In fact, almost all flooring can be made from recycled materials. The floor (and the outdoor deck) can be made from recycled barn-siding, recycled bowling alleys, or salvaged floor themselves.

Flooring That Helps Heat: In this renovation by CEDG Architects, a concrete slab floor functions as a thermal mass which is heated by the sun when the patio doors are open. (See Chapter 44.)

Chapter 5

The Roof:

New Options

Both the shape and materials used in roofs of eco-renovations are among the most important decisions to be made. While the shape of the house's roof is usually a traditional gable or flat surface, there are many new options, including small overhangs and levels added to the original roof line, as well as dramatic curving or pitched additions.

In fact, the shape of the roof is of great importance to the building's energy costs. Flat roofs lend themselves to roof gardens, which are both energy savers and aesthetically attractive options. A slanting roof, if it is south-facing, provides a site for solar panels and reflective surfaces. Another untraditional roof is the butterfly pavilion, in which a roof—made of fabric or other lightweight material, "floats" above the house or its porch, and is adjustable, depending of weather and climate. The striking curved roof allows winter winds to flow over it, as well as providing a setting for solar panels.

Traditional roof materials include several types of slate or tiles (which can last for as much as 75 years), and clay tiles, which are long lasting, and in light colors, can reduce the roof's surface temperature. Traditional shakes and shingles are an option and can be purchased with certification. The partly glass roof is another commonly used alternative, providing passive solar energy and brilliantly light interiors. And recycled materials are, of course ecologically sound solutions: shingles, tin, slate, and even old roadside signs have all been used effectively.

But newer materials are being used by eco-conscious builders too. Among them are metal and plastic roofing materials, with "cool roof" properties. Perhaps the most popular are the recent "white roofs." This is a cost cutter: a metal or plastic roof that both cuts energy use and helps cool the planet. We may all know that wearing light colors is a cooler option in hot weather; the same

The White Roof: In warmer climates the energy-saving white roof reflects back a large amount of the sun's rays, as in this Austin, Texas, example by Bercy Chen Architects; this house also features a roof-top space called a "brise-soleil." (See Chapter 3.)

principle applies to roofing. The white roof is, in fact either white or a pastel color, and as such absorbs less heat than a dark one. It may reduce air conditioning costs, and experts estimate that increasing the use of white roofs can cut carbon dioxide emissions (from air conditioning) dramatically. Several states have building codes that encourage the use of white roofs, as does the E.P.A. The white roof reflects back a large amount of the sun's radiation and is particularly effective in warmer climates.

The "living roof" is another option for those with flat surfaces to plant. The use of vegetation for aesthetic purposes is, of course, not new, but more recent innovations bring unexpected benefits. It has been known for the last decades that plantings on the roof (and the earth beneath them) are great insulators and improve the microclimate, while absorbing dust and pollutants. In addition, green roofs can collect rainwater. (Such collection systems can take in some 70,000 gallons of water in some climates—water that can be used for cooling, flushing, etc. A green roof can also combat variations in temperature. If added to a planned environment of plantings inside and out, the house can become a sustainable oasis. In addition, a living roof provides a habitat for birds and insects and protects the under-roof from damage and U.V. degradation. The green roof can be installed on pallets or trays of earth; although the earth may be only several inches deep, each pallet still weighs some 15 pounds per square foot.

The cost of installing these various types of ecologically sound roofs varies considerably. Using recycled materials is, of course, the least expensive option. A white roof is not necessarily more expensive than a traditional one. A green roof costs around \$4 or \$5 per square foot (plus the cost of reinforcing the roof to hold the earth itself.)

The Shape of the Roof : Extraordinary curved roof for this solar-powered Vermont home, designed by architect Ted Montgomery, encourages cold winds to flow right over it. (See Chapter 25.)

The Roof Pavilion: Designed by architect Rick Harlan Schneider, a fabric roof canopy called a Butterfly Pavilion floats above this townhouse; it can be adjusted to the climate. (See Chapter 16.) Photo by Dan Redmond.

Rooftop Gardens: More and more popular for their aesthetic and insulating effects, roof top gardens, like this one by architect Robert Renner, provide significant savings in energy. (See Chapter 4.)

Chapter 6

Energy Alternatives:

Power from the Sun and Wind

In the past few years alternative energy sources have become increasingly available for home use. Solar and wind power will provide "the energy of the future," according to designers, although there are still problems connected with home use, including costs. However, there are many different answers to alternative energy that are both affordable and sustainable.

Solar remodeling refers to two kinds of energy: passive and active. Passive solar energy is energy that depends on treated glass to retain the sun's energy and bring it into the house, and perhaps even to heat a slab of flooring. A storage system can retain some of that energy on dark days. Active solar energy describes a system of devices used to capture the sun's rays with photovoltaic cells and then store that energy for future use for heating, electricity, and hot water. These devices include roof panels, modules, and various systems to store and deliver energy to the home.

As there are so many types of systems, the renovator interested in solar solutions should research the field to find the appropriate one for the climate and style of the building. (There are many books and organizations that explain the technologies available.)

Many people opt for combinations of active and passive solar energy, or for combinations of solar and traditional technologies. Some use solar power only for hot water, along with traditional types of energy for heat and electricity. Some combine passive solar with gas or oil heat, or even wood stoves.

The cost of going solar is, of course, variable. Passive solar is simply the cost of the installation of various glass and reflective areas, and perhaps a storage system. The cost of active solar is much greater, but is an excellent investment over time; experts report that active solar pays for itself in fifteen years. Many local governments are offering tax breaks and other incentives.

A brief look at passive solar energy projects shows an emphasis on high performance windows with low E coatings that can act as solar heaters. South-facing windows, as well as skylights, use the sun's rays on sunny days, and rely on storage systems for other weather. A greenhouse (or even a windowed green hallway) — or any room with a large expanse of glass — can be converted for passive solar energy. Especially designed floor slabs that retain heat and are open to the sun's rays are also part of the passive technology.

The use of active solar power involves a system for collection and storage. Solar roof panels—the most common system for harnessing the sun's power—are usually placed on a south-facing roof, at an angle to catch the winter sun. (Some designers make the panels moveable; they pivot to follow the sun.) A series of devices carry the energy into the house, where it is stored and used for power. The average installation does not cover all of the needs for electricity, heating, or hot water. And most require a traditional back-up system. However, large amounts of energy can be provided by active solar technology, with a significant lowering of monthly bills. It is important to figure out exactly what power needs can be economically met with active solar energy and whether the climate is appropriate for active solar energy. (Keep in mind that areas with heavy snow may not be good matches for solar roof panels; the retained heat can melt the snow and cause too much moisture to permeate the roof.)

New options for solar energy are appearing all the time. Some are new building materials, so that the house does not even need solar panels. Among these new products are tiles and shingles that have photovoltaic cells sealed within them. They come in a variety of color (terra-cotta style in the Southwest, for example, or gray for New England's salt-boxes). They even come in a variety of shapes. (The cost is around $20,000 for one house, which is 10 or 20 percent more than solar panels may cost) but they are less obvious and can be added to a roof during a renovation.

Wind power is another alternative energy source. The homeowner obviously cannot set up a wind farm to generate his own power. But if we look back to old-fashioned farms, we see the individual windmill, a traditional part of the farm landscape, now returning to use (as can be seen in one of the mid-western projects in this book). There are a number of new inventions, as well.

For between $780 and $1250 the renovator can buy a miniature rooftop turbine, which purportedly produces 80% of a home's energy needs. Designed by Philippe Starck, it has transparent plastic blades and is virtually invisible up on the roof. Turbines on the roof can generate energy for a fraction of the price of commercial energy charges. Turbines are only one of the new inventions harnessing the power of the wind.

The new addition by Shelter Architects in back of a Minnesota home provides both passive light and heat with a huge expanse of high-performance glass. (See Chapter 10.)

A "greenhouse hallway" addition to a home in New Mexico by Edge Architects brings light and heat into the interior. (See Chapter 23.) Photo by John Robledo

Passive solar energy is created by South-facing, insulated glass wall with a vast roof overhang, in this Kentucky design by architect Mark O'Bryan. (See Chapter 11.)

A trombe wall in this canal house renovation, by John Ringel of Jersey Devil Architects, involves a glazing system that absorbs the sun's rays, with a heat-retaining masonry wall several feet behind the glass.

Active Solar Ideas

Roof-top solar panels are a popular choice for active solar energy in a small building renovation, as in this design by Maine architect, Richard Renner. (See Chapter 4.)

In this renovation by architect David Hertz, an array of 20 solar panels was placed on the roof of a large house, while other parts of the roof have flat-plate collectors for hot water and heating systems. (See Chapter 19.)

The solar umbrella designed by Pugh and Scarpa Architects provides thermal protection and 100% of the energy needs for this California house, with a series of solar panels wrapping around the southern elevation. (See Chapter 7.) Photo by Marvin Rand.

Instead of changing the traditional New England style of this shingled house, architect Peter Kurt Woerner placed a solar panel system (known as a solar corral) away from the building, and partially obscured it with a native stonewall. (See Chapter 2.)

Chapter 7

Insulation and the Thermal Envelope

There are a number of ways to save energy—none perhaps as important as good insulation. And there are many methods to complete the thermal envelope of the building with energy saving (both heating and cooling) in mind. Data shows that in the average American house some 40% of primary energy is lost. The new passive house is aimed at creating the air-tight thermal envelope which can save up to 90% of energy costs.

The climate you live in is the first consideration. This is a relevant subject because ecologically sound building takes into account the specific needs of a house in relation to its particular climate. The innovations shown in this book may concern water conservation in a dry area, or a tight thermal envelope with protection against ice and snow in a cold area, or ways to deal with heat and humidity in a hot climate.

The next consideration is the type of insulation—one that will require the minimum amount of energy for comfortable living in your climate zone. Most insulation is used within walls, roof, and flooring.

THERE ARE A NUMBER OF OPTIONS.:

- **Sprayed foam insulation** can be either made with polyurethane, petroleum, or even soybeans. It is sprayed into walls, using holes in the preexisting exterior or interior walls, or into new wall cavities.
- **Cellulose insulation** is well regarded by eco-builders, both because it is made from recycled materials (like old newspapers) and because it can be blown into existing walls. **Cotton insulation** is another recycled material; it is commonly made from old fabrics like denim that are treated for bug and moisture resistance and blown in.
- **Fiberglass insulation** is probably the most common form of insulation used in the United States. Precut, flexible batting and rolls of fiberglass come in many shapes and sizes, has a vapor barrier on one side, and can be fit into various spaces. Fiberglass insulation can be purchased formaldehyde-free.
- **Icynene insulation** is another option; it expands within a few seconds to fill the building cavities.
- **Radiant barriers** are panels that help reduce heat loss.
- **Structural Insulated Panels (SIPs)** are a composite form of rigid insulation. They consist of two layers of structural board with an insulating layer of foam in the middle; they are so strong that they can replace some traditional building elements.
- **Frost walls** are a form of reinforced concrete wall. This is actually an extra wall, built low outside of the foundation to protect the building damaging freezes.
- **Trombe walls** are one of the earliest and most effective eco-systems for insulating a home with a south-facing side. Construction involves a glazing system, which absorbs the sun's rays; there is an inner—perhaps masonry—wall several feet behind the glass. As the interior wall receives the heat from the outer glass, it radiates warmth into the interior of the house.

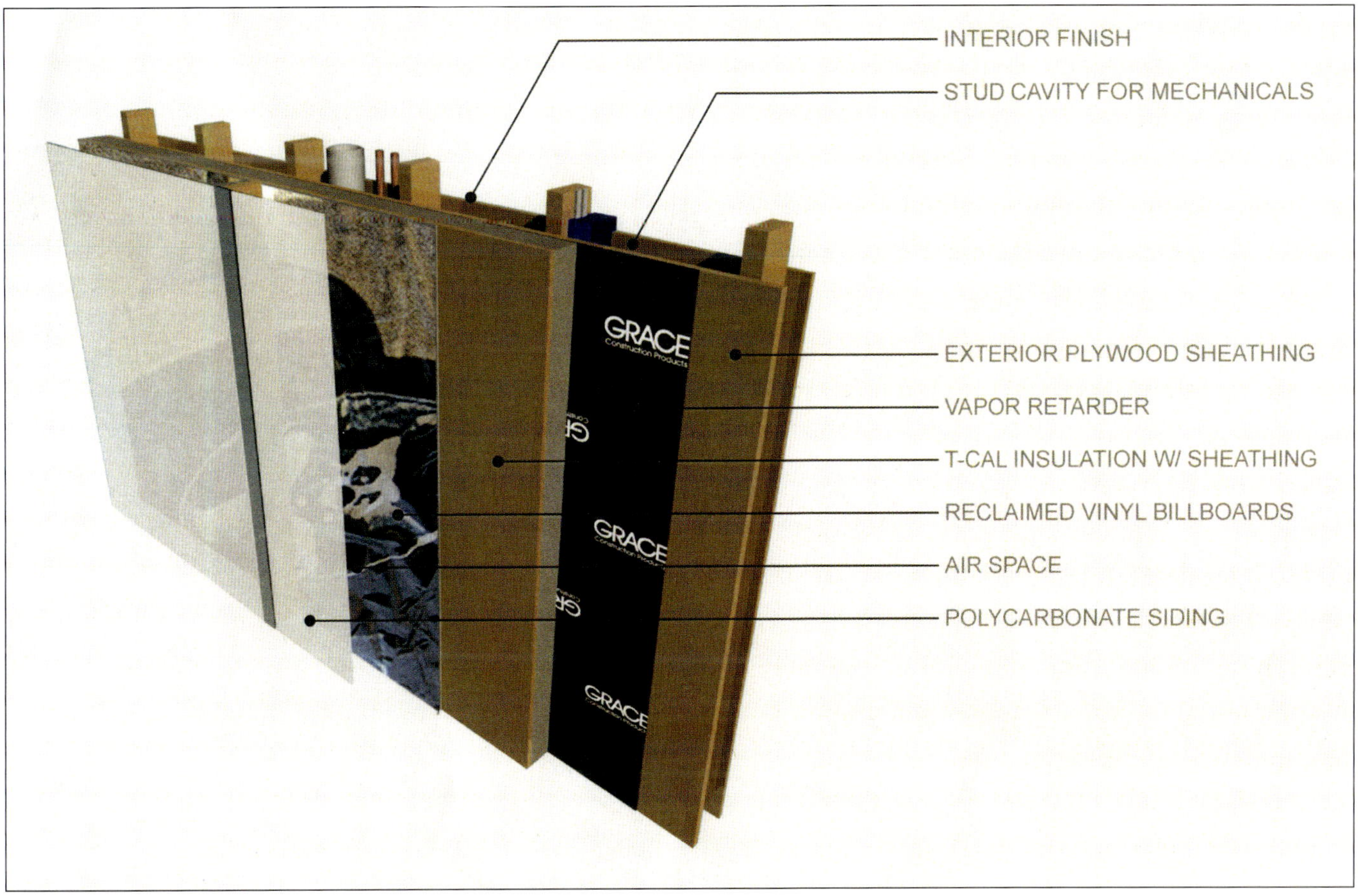

Tight Thermal Package: Diagram for a tight thermal envelope by Locus Architects shows layers of the house in cold Minnesota climate. (See Chapter 30.)

Rigid Foam Insulation: A picture taken during construction shows a New Mexico house renovated by Edge Architects with adobe brick and stucco, as well as 2 inches of rigid foam insulation. (See chapter 6.)

Wrap-around Insulation: Before and after photos show how Architect Robert Knight removed the shingles in this Maine seaside house, wrapped the building with an air barrier, and reshingled the house. (See Chapter 22.)

Chapter 8

Geothermal & Natural Solutions to Insulating the House

One very old technique that has reappeared in the quest for ecological building solutions is the use of the earth itself as an insulator. Long used by rural builders, utilizing the earth has become an attractive option.

Among the most innovating solutions to the use of energy and insulation are geo-thermal systems. The earth itself is an amazing saver of heat and new technologies are using ground-source heat pumps and a variety of other methods for saving energy and keeping the house insulated.

New ways to generate energy from the earth are still being developed, but this technology is fast becoming the choice of architects and builders. There is a relatively constant temperature just below the earth's surface (some six feet or so deep). A geo-thermal heat pump draws the warm air into the house in winter and pulls it out in warmer weather.

Experts predict a saving of up to 65% on energy costs with a geothermal pump. However, adding such a system can add thirty to forty thousand dollars to the cost of the project, according to one architect, and there are many unknown factors involved, like the depth needed in your area.

A much simpler way to use the earth for insulation is the berm. The earth surrounding a house can be used as a buffer against the elements (in a technique that calls to mind pioneer sod houses). The piling of earth against a wall at the side of a house is considered an excellent form of insulation (and it also cuts down on noise). Earth-bermed architecture loses less heat than other buildings; the earth retains a 55 degree temperature. (The downside to berms is that they may bring in moisture, which condenses on cooler surfaces, but additional insulation can remedy this problem.)

Trees and gardens are not just for scenery any more. Today they help the ecologically-oriented control the micro-climate and the energy needs of the house. The roof garden both combats variations in temperature with a layer of earth atop the roof's surface, but it is also an attractive addition to many buildings, in addition to being a saver of energy. Although several inches of earth atop a building or a deck can be heavy and need extra support, the overall insulation provided by such a garden is an attractive option.

Similarly, the green wall is another new method of using nature for insulation. The vertical green wall has plants growing up the side of a building to create a "growing wall." In fact, the green wall is seen by many as a bold step toward improving insulation and cutting costs (it is used more widely in Western Europe). However, in downtown Portland, Oregon, even a Federal building will soon have just that: a 200 foot high wall of plants, using some 60% less energy within, and saving money and water as well (with the use of salvaged rainwater).

Earth Berm: Vermont architect Ted Montgomery used an earth berm (a hill of earth against the house) to protect and insulate the house from changes in temperature. (See Chapter 8.)

Chapter 9

Ventilation:

Bringing in Fresh Air

One effect of the tight thermal envelope that is the aim of every energy-conscious designer is the need for ventilation. Just as there are numerous ways to close up a house to cold air, there are many possible ways to bring fresh air inside.

Understanding how ventilation works is key: hot indoor air has to be replaced by cooler outdoor air. Hot air rises; the "chimney effect" is created when you let in air through windows at a lower level to rise to some kind of operable windows, skylights, or other apertures at the top of the house. Fans—particularly those that blow outwards—can help with this process. Cross ventilation is also essential; air exchange from window to window helps create airflow.

The siting of the house or addition is the architect's first consideration in planning good ventilation. Prevailing winds, shading trees, and southern sun are all taken into account. Some plans call for a "dog trot" (an open-ended corridor to bring breezes through) or a breezeway. These built-in designs encourage fresh airflow. The placing of windows where they are most likely to ventilate the house is another basic choice.

There are many new and innovative ways to create ventilation as well, including solar fans, cooling towers, heat exchangers, ventilating chimneys, various types of window fans, and combinations of these devices. All of these plans aim to make air conditioning (and its power drain) unnecessary.

Some of these systems, such as the heat exchanger, depend on re-using the stale air from the interior of the house. The object of these methods is to avoid opening the house to drafts, while still providing ventilation.

Another design for air circulation is the cooling tower, a device which sends heat into the out-of-doors by cooling with a stream of water. It draws out hot air in warm weather and circulates warm air in the winter. This system is fitted with an overhead fan in the center of a tower at the top of the house.

Ventilation chimneys are also an option; this passive device exhausts warm air from the highest point in the house; insulated hatches open and close thermostatically, with cool air introduced through floor registers.

The most common ventilation plans, of course, involve the simple opening of windows. But here too, new methods are able to regulate airflow, using a variety of pulleys, hatches, thermostats, and fans.

Regulating Air Flow: An ingenious system of flip-up windows provides ventilation in this Alabama project by Taylor Dawson, (See Chapter 20.) Photo by Robert Fouts.

Air from Above: High, operable clerestory windows and skylights are used by architect Angela Davis for passive convective ventilation. (See Chapter 27.)

Ventilation Chimneys: In this California house designed by Obie Bowman, ventilation chimneys bleed off summer heat build-up. (See Chapter 39.) Photo by Tom Rider.

Chapter 10

Choosing Materials:

The New, the Reclaimed, the Sustainable

The eco-conscious renovation will bring up a great many choices regarding the materials used. From siding to piping, flooring to roofs, countertops to wall coverings—new and sustainable materials are available to the discerning renovator. There are even a variety of materials to be salvaged or reused in a new way. When planning a renovation, the designer should not overlook the availability of materials. Keep in mind that the carbon footprint left behind in long trucking distances, as well as the expense of using materials that are not native to the area. Just as builders in the southwest have long used adobe brick—made from local clay—and those in New England—native stone—the choice of materials should be carefully weighed with both distance traveled and sustainability considered. Sand used in the making of glass and clay for bricks are also easily replenished materials. A brief description of a few of the most popular, ecologically sound options follows.

- **Recycling:** Certainly the most ecologically sound materials are those that have been used before: the recycled choice. There are numerous places to find pre-used building materials, ranging from basic wood to old mantels and fine windows. And then there are the imaginative reusing of materials: among the most unusual materials for building such as old billboards, or re-configured shipping containers. (The Building Materials Reuse Association (BMRA) has listings by state of places to find such materials.)
- **Wood:** Many architects are choosing woods that are certified to be easily, and comparatively quickly, replenished. Certain woods, like mahogany, take a long time to grow and a forest cut down not only depletes the supply but diminishes the atmosphere around it. Among the choices of the eco-builder are teak, ipe, bamboo (actually a grass but used as a form of wood), Douglas fir, and a new sustainable forest wood called Ovangkel. In each area of the country certain woods are prevalent (such as redwood in California, Pecan wood in Texas, or fir in Maine), which the eco-conscious designer and builder may choose for beams, framing studs, flooring, etc., to avoid shipping from great distances. (It is not hard to determine where a wood comes from; organizations including Green Seal and the Forest Stewardship Council (FSC) indicate on their seals that the lumber has been properly harvested.)

Found Objects for Building: A most unusual translucent wall, in this Minnesota house by Locus Architects, is made of discarded vinyl billboards. (See Chapter 30.)

- **Alternative and manufactured woods:** Another option to enable the builder to avoid using hard woods that may take a quarter of a century to grow, are alternative and manufactured woods. For example, tree bark—like that used for cork flooring—is a tree product that is now much in use for its softness and sustainability. Manufactured logs can be made from precast concrete in the shape of tree trunks; they are painted to look like real logs. Sturdier and more eco-friendly that the real thing, such manufactured products may appeal to eco-conscious builders. Other such "logs" are made of pine veneer on top of hardboard, over foam.

Metal as Building Material: A new addition to a Texas house, by Webber + Studio, is wrapped in metal cladding, with a metal roof as well. (See Chapter 17.) Photo by Tom McConnell.

- **Concrete:** Insulated concrete forms can be used in hollow blocks or panels or even planks, to form walls and flooring. There are various types of cement and concrete. Popular for their thermal properties, such materials as "syndecrete" can be obtained in non-toxic forms and used as a slab for solar energy, radiant heat or even elegant flooring.
- **Glass:** Triple-glazed, low-E window glass is a common choice, as is glass block. Both are easily manufactured and sustainable options, particularly apt for those considering passive solar power.
- **Rammed earth:** This is another material now much in use, although it is an age-old technique. It is a mixture of earth, sand, and cement that is tamped into a preshaped form, and then allowed to harden and reinforced with steel.
- **Stone and brick:** Popular alternatives to wood houses are stone, particularly in areas like New England where it is so common, and, of course, brick (made from local clay). Used for major construction, as well as for fireplaces, bathroom and kitchen floors, stone is often recycled or combined with other materials. Even stone is now engineered into "composites;" for example, combinations of quartz and polyester resin can be an alternative to natural stone. Adobe bricks, made of clay and straw, are noted for their insulating properties.
- **Plastics:** Plastic piping is usually a polymer made from petroleum. It is considered an acceptable alternative to copper piping. Materials other than plastics are recommended for countertops; among the most sustainable are concrete, a natural stone (such as marble), mixtures of paper and resins, glass composites, and various types of tile (made from ceramics, porcelain, or resin).

Manufactured Building Material: This design, by architect David Hertz, used a recently developed type of concrete called "syndecrete," which is made of 41% recycled content; it absorbs heat and functions as a "solar sink." (See chapter 19.)

- **Straw-Bale:** An old, but venerable material for building is the straw bale.. Straw bales, are of course, easy to find in agricultural areas; each bale is thick and can be piled up to create well-insulated walls. Straw bale building meets eco-criteria: it is local to many communities, no trees are cut down, and it is a good insulator and energy saver, and is extremely economical. Bales of about a foot and half high and three feet wide are stacked atop one another in neat rows. They are supported by wood stud framing and subsequently covered with stucco or other material for the exterior, and plaster or wallboard within the interior. However, it should be noted that many architects avoid straw bale building because the straw can retain moisture and become moldy if it is not thoroughly dry before enclosure. They maintain that it many require three coats of stucco to avoid dampness.

Straw Bale Building: An inexpensive and insulating material, straw bales have lumber framing and an outside surface of stucco in this design by John Ringel of Jersey Devil. (See Chapter 34.) photo by Shekinah Rae.

Chapter 11

Glass and Its Surprising Properties

One of the easiest renovations to make to a home is the type of glass used for windows, skylights, doors, etc. High performance glass is now widely available.

The thermal elements of glass are as important as any part of the "envelope" that keeps the house warm in winter and cool in summer. Traditional glass can account for vast amounts of heat loss and gain—accounting for as much as a third of the interior's heat or cooling.

Energy efficient windows are characterized by double—or even triple-depth; they have several panes with a "dead space" between them. Such double or triple glazing makes the windows at least twice as energy efficient.

The space between the panes of glass in double or triple depth windows can be filled with argon or krypton, which adds to the insulating properties. In addition, foam, rubber, or metal can be used to separate the layers of glazing.

Of particular importance to the new windows is the coating used on the glass itself. Low E (emission) glass is glass that is coated; it has a transparent metal film that makes the glass less susceptible to heat and cold. When facing the heated side of the window it reduces the warmth coming through the glass. Thus it reduces heat loss in a cold climate and protects against heat in warm climates. Similarly, thermopane windows prevent heat loss and condensation in colder areas.

When choosing glass for the renovation, the buyer should look for the NFRC number, which will indicate its performance rating and its energy efficiency. (See EnergyStar.gov.)

Placement and orientation of windows is of major concern to the energy-conscious. About twenty percent of electricity used is for light, so a well-placed window can provide passive solar energy, and cut down on the use of electricity.

Skylights bring in light from the most direct angle of the sun. Recent innovations include raised skylights that collect the sunlight above the roof level, and transfer it through an aluminum, mirror-finished pipe, directly to a diffusing lens in the ceiling.

Glass block has been used for decades (it was particularly popular in the Art Deco era); it has recently come to be seen as a particularly effective and sustainable material. Glass block is part window, part wall, and it can be used in a variety of ways, creating interior divisions, as well as bringing in light while protecting privacy. The first environmental benefit to the use of glass block is its ability to bring in natural light and reduce the need for electricity. Even the thickest glass block still brings in some fifty percent of daylight (and thinner versions can bring in up to eighty percent). It also has greater insulating power than regular glass. Made of silica, sand, limestone and soda ash—all combined and heated within two halves of glass, the blocks have a decorative, wavy design. Environmentalists appreciate glass block's low pollution, easy manufacture, and the fact that it can be recycled.

Reused Garage Door: A huge, recycled glass and aluminum garage door brings light into converted Chicago power station, as designed by Ann Temple and Ken Schroeder of SMNGA. (See Chapter 32.)

Internal Window Well: The window bank in this Minnesota house by Shelter Architects has a colored internal window well to "color the light and warm the space." (See Chapter 10.)

Pyramidal Skylight: In Obie Bowman's design, a pyramidal skylight provides light into the center of the house through perforated metal screens. (See chapter 14.) Photo by Tom Rider.

Glass Block Floor: In this house by John Ringel of Jersey Devil, glass block in the second story floor allows light from above to reach the floor below. (See Chapter 5.)

Glass Block: Recycled glass block covers the front façade of a California house by Bercy Chen Architects, both bringing in light and providing insulation. (See Chapter 29.)

Chapter 12

When Water is Scarce

There are a number of increasingly popular methods for saving water—particularly in areas where water conservation is not just a slogan or a new idea, but a necessity. Either by saving rainwater or reusing household water, these technologies can be effective ways of going green.

One such system is known as a roof catchment system. It is simple in concept; for some 4000 years people have been saving rainwater by collecting it from roofs. There are many easy ways to collect rainwater, including the usual plastic downspout, as well as the rain tube, the rain barrel, the rain chain, and the rain screen.

A roof catchment system consists of three parts: a unit that collects the rainwater, pipes that carry it elsewhere, and another vessel that stores it for future use.

The rain tube (which eliminates the problems with clogged downspouts) is placed inside the gutters and allows the free flow of rainwater into barrels placed below.

The rain chain, is an old Japanese technique, in which a copper, aluminum or iron chain (with or without decorative features) is suspended from the roof. As an alternative to the usual plastic downspout, the rain chain guides the water down to a catchment barrel for future use.

The rain screen, like the rain chain, is set up to carry rainwater via the use of slats down to waiting receptacles. It can add a picturesque design element to the house.

"Gray-water" reuse systems are effective ways for saving, in which contaminated water is "cleaned" by the use of microbes and bacteria that break down and destroy contaminants. This method involves the use of a septic tank, set well away from the house.

Rainwater Recycling: In this stunning redo by Bercy Chen Architects, an extensive rainwater recycling system is both ecologically helpful and aesthetically pleasing. (See Chapter 29.)

Rain Chain: An old Japanese technique called a rain chain draws rainwater downward from the rooftop in this design by Rick Harlan Schneider. (See Chapter 16.) Photo by Dan Redmond.

Storm Water Retention System: In this house renovation designed by Pugh and Scarpa, a storm water retention system functions along with drought-tolerant xeriscape garden. (See Chapter 7.)

Chapter 13

Growing Things:

Nature as Part of the Plan

One of the most eye-catching—as well as ecologically sound—elements of green building is the use of systems that are, indeed, green. Buildings with vegetative roofs and facades, indoor plantings, and all kinds of greenery on decks and patios are immensely helpful to the environment and to the thermal envelope of the house—as well as being a pleasure to look at. What is known as biophiliac design is a recent return to what was once more common: the use of plantings for insulation, food, camouflage, and protection.

Roof gardens can contribute tremendously to the thermal envelope, with their layers of earth insulating the roof. Vegetative facades can be seen on tall city buildings around the world, while on a more modest level, architects are also using plant-covered walls for insulation and design in private houses.

Trees and shrubs are among the best methods for moderating wind. They also filter noise, and pollution, reduce glare and block the sun in hot weather, sometimes forming a canopy that pushes breezes to ground level. In winter, deciduous trees let warming sun through, while evergreens form protective weather blocks.

Nature's Help: Two ways in which architect Ted Montgomery uses plantings in his Vermont house: a carefully organized roof garden (above) and an indoor garden room (opposite, top).

Creating a Micro-climate: Architect Angela Dean creates a micro-climate to help insulate this Utah house, with plantings on many different levels and surfaces. (See Chapter 9.)

Chapter 14

Choosing Paints and Finishes

Since the green movement has caught on, manufacturers are producing healthier paints and finishes (called LOW-VOC and NO-VOC) in a wide range of colors and surfaces. VOC stands for volatile organic compounds, such as certain chemicals and lead (which was banned in the United States in 1978). VOCS are usually added to paint to make it dry faster and go on smoothly. While traditional paints may flake with age, releasing tiny but toxic amounts of gases into the air, these new paints and finishes are considered much better for public health and the environment. (In using LOW VOC paints, note that the amounts of toxic chemicals listed may apply only to the base, not the colorant, but some brands are available that limit them for the colorant too.) Primers too can be non-toxic. The use of low or no VOC paints is an easy and not very expensive alternative to regular paint, and is highly recommended by our architects.

"Natural paints" (or NO-VOC) are those made from plants (including citrus fruits, seed oil, and resin) or minerals rather than chemicals or petroleum. They are biodegradable, have little toxicity, and reduce the environmental impact of our usual synthetic paints. There are various kinds of low or no VOC paints; you can read can labels at the paint store to find those with the least (or no) amount of such chemicals as methylene, chloride, tolvene, xylene, acetone, and other petroleum-based products. There are also Material Safety Data Sheets (MSDS), which are supplied by the makers; they list the amounts of chemicals in the mix. Among the natural types of paint are some very unusual new mixes: one kind is made with milk, combined with mineral pigments and lime. The soft, somewhat chalky finish is popular for unfinished wood, but it cannot be used in damp areas.

Off-gassing is how toxic elements enter the home—from paints, stains, adhesives, carpeting, and insulation. Off-gassing can continue for years after the material is put into the building, and today's airtight thermal envelopes mean that there is little fresh air to mitigate them. Therefore, it is recommended that you choose non-toxic paints and finishes from the start. Experts say that the worst of the air pollution caused by paints and finishes occur before they dry, so be sure to leave the windows open!

Finishes—both those that come with building materials, and those that are added later on—also often have toxic qualities. But alternatives are available; read the ingredients carefully. Some non-toxic wood finishes are made from vegetable oil and waxes; others are water-based acrylic.

Non-polluting Sealants: In this project by Rick Harlan Schneider, cedar siding and ipe decking are sealed with non-polluting finishes: the cedar with low-VOC waterborne sealer, and the ipe flooring with natural tung oil. (See Chapter 37.) Photo by Dan Redmond.

Formaldehyde-free: In this redo by Richard Renner, finishes on cabinets and flooring are formaldehyde-free. (See Chapter 41.)

This kitchen renovation by Leger Wanaselja Architects uses no-VOC paints, natural, oil-based wood finishes, low-VOC caulk and adhesives. (See Chapter 41.) Photo by Emily Hagopian/Robert Kolderup.

PART II

45 GREEN SOLUTIONS: Architects Explore Eco Renovation

1

The Greening of a Modernist Gem:

Preservation and Innovation

Tigerman McCurry Architects, Ltd.

Photos by Steve Hall, Hedrich Blessing Photographers

Designed in 1953 by a well known architect, Richard Barancik, this modernist house was in need of a green update. But preservation of the original design was of paramount importance to the new owners and all changes had to be made with the aesthetic of the 1950s one-story, flat-roofed house kept intact.

The goal was to "utilize the language of rough stone, expanses of glass, and elegant wood paneling that characterized the original," said the architects, Tigerman McCurry, Ltd. of Chicago.

But the house needed increased energy efficiency, an enlarged kitchen and bedroom, as well as a redesign of its surroundings. The total new footprint was just over 1000 square feet (added to the 5000 square feet of the house before renovation).

The original house—considered radical in its day—was, in fact, ahead of its time in having radiant heated stone floors and deep overhangs for shade. But there was much to be done for true thermal efficiency, and the careful and calculated renovation that would preserve the modernist look of the house included many eco-elements while maintaining the mid-century modernism of the original design.

One of the major elements of the retrofit was the matching of materials. Tennessee Crab Orchard Limestone, which had been used for floors throughout the house, was still available at the same quarry. Craftsmen duplicated the original layout. The wood paneling—a central element in the interior—had been made of Paldao wood—a rare commodity from the rain forest. New paneling was made from Ovangkol, a sustainable wood, which was milled and stained to match the original. New wood windows designed to match the originals stylistically and similar bronze hardware replaced the old. The kitchen floor was redone in terrazzo, custom blended to match the original.

Other major updates to the house included an energy retrofit that extended the radiant floor heating system, insulated glazing, new eco-friendly finishes, cork flooring, and ceramic tile work.

The outdoor setting also got a makeover with the environmental and aesthetic benefits of more trees and plantings (designed by Maria Smithburg of Artemesia). The asphalt driveway was replaced with the more environmentally suitable drainage system of pea gravel. An elevated terrace, a pool, and patio areas were added.

The redone house is now a spectacular example of mid-twentieth century design, but with a twenty-first century greening.

Before renovation: The original 1953 modernist design was to be preserved despite changes.

After renovation: The exterior of the house with additional greenery. The hard surfaces were replaced with landscaping.

Kitchen before and after: In the redo the original rare wood (Paldao) was supplemented with new, sustainable wood, carefully milled and stained to match the original in the expanded kitchen and breakfast nook.

Craftsmen duplicated the original layout using limestone from the original quarry in Tennessee, with terrazzo added to complement the limestone.

L-shaped, one-level modernist house was opened up to an elevated terrace and pool, and patio area.

2

The Retrofitting of an 18th Century Island House

Peter Kurt Woerner and Associates

On Rhode Island's rural, scenic Block Island, a beautiful and traditional 18th century house got an ecologically innovative make-over by architect Peter Kurt Woerner. The steps he took to make the house both more "green" and more economical included many of the most successful new strategies—from solar technology to the salvaging of materials.

A sampling of these techniques includes several related to the island location of the house: hurricane–proof building, local, easy-to-get products, and alternative energy. New methods of insulation were another important consideration for a house in this exposed, windy setting so near to the ocean.

The most striking addition to the house is the solar panel system located below the house (well away from the spectacular ocean view). Known as a "solar corral," this area is enclosed by stone walls. It provides electricity for most of the year (island electricity rates are exceptionally high), and a solar hot water panel system provides most of the hot water from late spring to early fall, and close to half during the cold winter months. The solar set-up also creates enough heat for hot water radiant floor heating from a hot water boiler.

Insulation was a primary consideration. A combination of fiberglass insulation and poly-isocyuranate foam with a vapor barrier was used in the walls and the ceilings to create a tightly insulated interior. The new energy-efficient windows were insulated as well, with low-E glazing.

Among the local products were native Rhode Island stone for the chimney and stone from Block Island itself for outdoor walls. And many elements were recycled: salvaged lumber, original floor beams and joists—even the antique 19th century slate sink—were all salvaged, adding to the economical and green footprint of the house. Except for the partially hidden solar panel system, it is hard to see the many sustainable features in this seemingly traditional New England home in its spectacular island setting.

The Block Island house before renovation shows its rural setting exposed to the elements.

The completed renovation shows traditional island home with solar panel system partially hidden by native stone walls. It provides a good part of the electricity and heat to the house.

An interior view shows original pine boards for the second floor and new energy-efficient windows.

The kitchen reused flooring beams and joists for second floor framing, as well as the original 19th century sink. The wall outdoors is made of native Block Island stone.

3
Light and Shadow and a Brise Soleil

Bercy Chen Studio LP, Architects

Turning a traditional 1980s sort-of-Colonial into an energy-saving contemporary is not easy. But the location on the picturesque banks of Lake Austin—with 180 feet of lake frontage—made the owners decide to renovate with the latest in technology and design in mind. (With similar lakefront properties on the market for two million dollars and up, the cost of about $500,000 for the renovation seemed a comparatively good choice.)

The design featured a north-south orientation to accommodate great lake views. But because the house also faced a harsh western sun, extensive shading strategies were used, including 8-foot overhangs, automated external fabric window shades (which reduce 85% of the heat gain), and one of the latest in roof innovations. This is a brise soleil, a usable outdoor roof space, which provides shade for a new solarium on the ground floor.

The 1980s colonial-style Texas house had dormer windows and a traditional look before renovation.

in a Texas Lakefront House

The new, 21st century edition includes a solar roof and a reflecting pool.

Reflected light from a pond next to the house transmits light deep into the interior, and the new design opens up the top floor by transforming the dark attic space into a new live/play/study area for the children. The previously dark attic space was converted into a contemporary setting with two bedrooms. It has its own roof deck. Dormer windows were replaced by large areas of insulated glass that were sheathed inside with translucent panels—some bright red, some white.

A series of green roofs was designed to reduce energy costs and cut water run-off. Icynene insulation for better air sealing (in addition to sealing out dust and pollutants) was used in retrofitting the attic area. (Icynene expands to fill building cavities, reaching some sixty times its original volume in eight seconds.)

The outcome of this extensive renovation was the transformation of a traditional house into an energy-conserving and stylish new one

Pool abuts new glass solarium with wide expanses of insulating glass.

Reflected light from water and glass is mitigated by 8-foot overhangs and automated external shades.

Retrofitting of the attic into a live/study/play space for the children involved turning dark storage space into a contemporary setting with two bedrooms and its own roof deck. Red and white translucent panels transformed the original dormer spaces

4

Creating a Green Live/Work Loft

Richard Renner Architects

In downtown Portland, Maine, a small, nondescript commercial building became a live/work loft for the architect, Richard Renner, and his wife, a designer—as well as a prize-winning example of green renovation. The first LEED Home Platinum Rehab in the United States, this project focused on tight energy conservation, green materials, and salvage.

Previously a vintage clothing store with an apartment upstairs, the brick building was thoroughly redesigned; the first floor (just below street level) was to become the work space, with the second level converted into a loft-like living area with an interior balcony for additional work space.

The original brick building was hardly airtight. The old façade was preserved by the addition of new interior stud walls. The space between old and new walls was filled with closed-cell spray foam insulation. (The same insulation was added to the roof.) Triple-glazed, fiberglass-framed windows help make the new thermal envelope tight and bright, with a roof monitor to allow extra light into the loft living space.

Energy saving was a major consideration; it is now provided by a solar, grid-tied collector on the roof, a gas-fired furnace, a heat recovery and ventilation system, and an on-demand water heater. There is radiant heating (with two separate zones). The solar panel in spring and summer provides at least half of the loft's electrical needs. The new heating system has reduced heating costs to less than $400 during a cold Maine winter.

In addition, a roof garden and two green canopies planted with seedum and other succulents help insulate the building and reduce storm water runoff. The plantings were in two-foot square trays that could be arranged to suit the layout.

Energy saving is not the only green aspect of this refitting. The materials too were carefully chosen with many recycled and non-toxic choices: certified birch flooring, formaldehyde-free bamboo plywood for cabinets, countertops made of "Paper Stone" (a mixture of cellulose fiber and non-petroleum resin), recycled content tile, recycled glass tile for the back splash, salvaged steel railings and doors—these are all part of the eco-conscious renovation.

Among the other ecologically conscious decisions made by the architect was concern for the demolition materials; all left-over materials were taken to a municipal facility for possible reuse.

Before and after: This nondescript, downtown brick commercial building was converted into a live/work loft, receiving a major award for eco-conscious renovation. Note the solar panel on the roof of the redone building (opposite).

Photo courtesy of James Salomon.

Photo courtesy of Peter Vanderwarker.

Cabinets, flooring, and other woods are formaldehyde-free, with birch and bamboo as primary choices.

Photo courtesy of James Salomon.

A roof monitor allows light into loft living area; the office just above is bordered with perforated steel panels and hand rails.

A work area on first floor opens to the street; recycled-content tile covers the floor.

The roof garden helps insulate the building, and, with planted seedum, reduces storm water runoff.

5

Opening Up an Interior with Natural Light

John Ringel Architect, Jersey Devil

More natural daylight was one aim of the owners of this suburban house. They also needed an addition that would blend in with their neighborhood. They wanted to add a sitting room/sunroom off the kitchen, as well as a master bedroom and exercise room on the second floor. Their concern was for the new addition to be as transparent as possible, so that light entering it would reach the original kitchen. They also wanted to reduce their heating costs with solar energy.

The remarkable addition designed by John Ringel of Jersey Devil looks unassuming from the exterior, though it does have many tall windows. But the design gives little hint of the soaring two-story space within and flooding natural light that illuminates it.

Among the innovative ideas for bringing daylight into the interior is the use of glass block. (As we have noted, glass block—once a popular building material—has made a comeback; it is an economical and attractive way to create both privacy and light.) There are sections of glass block set in the walls to let light into the master bedroom and, of particular interest, placed in the upstairs floor of the new exercise room to illuminate downstairs space.

The open area of the addition, with its balcony and unusual design, creates an untraditional interior within a seamlessly blended addition. The combination of solar energy and natural lighting makes this an economical addition, while providing new living space for the occupants.

A two-story solar addition (facing south) includes a sunroom and exercise room below, and a new master bedroom above.

and Soaring Space

The interior of the new addition is designed with balcony, tall windows, and soaring space for maximum natural light.

The aim of the two-story open design is to create an area as transparent as possible. Note the light flooding from windows and the use of glass block on the right wall and in the floor of the second story, bringing daylight to the room below.

6

The Update of a New Mexico Adobe House

Ken Anderson and Pamela Freund, Edge Architects

Photos by Ken Anderson, Jamie Flynn, and Pamela Freund

This unusual renovation uses the traditional building blocks of the Southwest: adobe bricks. Located at the base of Taos Mountain, the house that Edge Architects undertook to convert was a 1960s structure, originally built for movie star Ali Ghito. Their aim was to turn it into an energy efficient and more spacious home.

Adobe is both a local and natural building material that "serves as an ideal thermal mass when coupled with passive solar design," say the architects. (Made of a mixture of native clay and straw, adobe brick is noted for its insulating and thermal storage qualities; it has been a traditional building material in the area for generations.)

The plan was both to update the original structure and to make an addition to it. A new roofed patio between the old and new structures has an operable glass wall, which allows the house

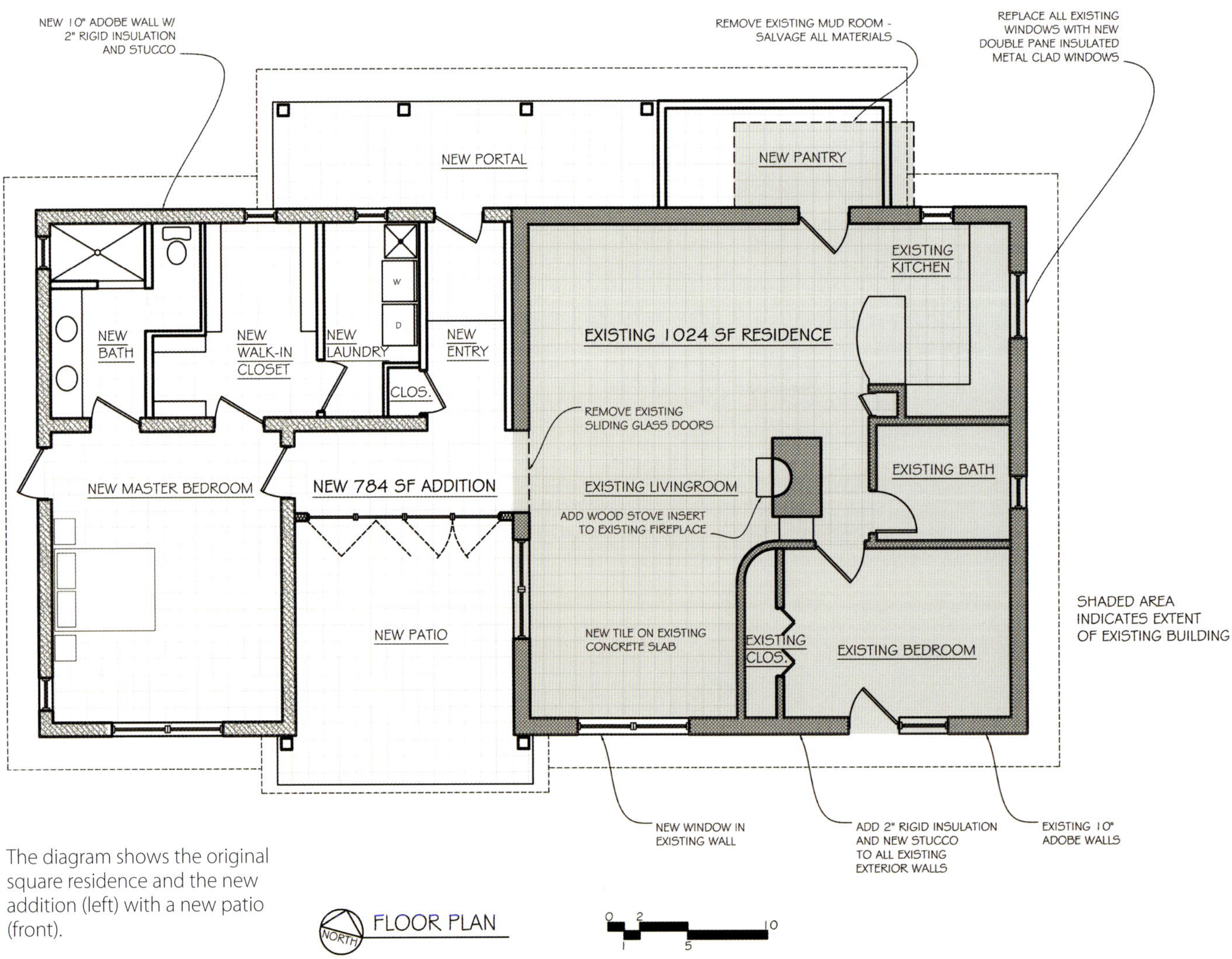

The diagram shows the original square residence and the new addition (left) with a new patio (front).

to be opened to the out-of-doors in summer. The architects added two inches of rigid insulation to the old walls and made the new addition of adobe and stucco equally well insulated.

The new design called for metal roofing with heavy insulation and sun tube skylights. Solar panels (installed behind the house) and a photovoltaic system provide electricity and hot water, as well as radiant floor heat. All of the glass windows, including large south-facing living room windows, were insulated and metal-clad.

In this dry area, the conservation of water is a major consideration; the architects designed a gray-water re-use system and a roof catchment system to provide water for the surrounding landscape.

An artist's rendering shows a reworked adobe house in view of Taos Mountain. It has a new insulated metal roof.

Solar panels and a photo-voltaic system provide hot water and electricity.

This picture, taken during renovation construction, shows walls with adobe brick and stucco; two inches of rigid insulation were added.

The new roofed patio between existing house and new addition is accessed by an operable glass wall, to allow house to be opened to the out-of-doors in summer.

7

Original Solutions:

The Solar Umbrella

Angela Brooks and Lawrence Scarpa,
Pugh + Scarpa Architects
Photos by Marvin Rand

Although it was inspired by the 1953 Umbrella House designed by noted architect Paul Rudolph, this answer to today's sustainable architecture takes solar energy to a new level. Working with a single story, 650 square foot bungalow in Venice, California, the architects at Pugh + Scarpa used a solar umbrella to transform the house into a contemporary and ecologically sound building.

The solar umbrella is a kind of canopy—a modern strategy that provides thermal protection in a climate of intense exposure. The site became 100% energy neutral with the use of passive and active solar technology. Changes included shifting the design toward the south, to create both more exposure to the sun (and the solar components) and a more gracious entrance. "A bold display of solar panels wrapping around the south elevation and roof becomes the defining formal expression of the residence," write the architects. Not only do the solar panels form a skin that absorbs and transforms the strong coastal sunlight for usable energy, but it also screens the body of the building from thermal heat gain and direct exposure to the hot, southern California sun. Solar-powered in-floor radiant heat is also provided.

The renovation also included enlarging the bungalow to 1900 square feet, creating a sizable addition to the south, and increasing the natural light, with the addition of stepped roofs, clerestory windows, and glazed walls. A floating steel staircase leads to the second floor and a covered patio. The ventilation system is also carefully calibrated; windows and an open stairway bring hot air up and through the house.

A major design element was the linking of indoor and outdoor spaces. The addition was conceived in relation to the landscape; a concrete pool defines the entry and cascades into a lower tier of water "that penetrates and interlocks with the geometry and form" of the house. Decomposed granite and gravel hardscape, and a storm-water retention basin are used instead of concrete or stone, allowing the ground to absorb some of the storm water, and to mitigate urban run-off into the Pacific Ocean nearby. The landscape around the house includes drought-tolerant xeriscaping.

Materials used throughout the renovation include all kinds of ecologically sound products: Among the many recycled materials are "homosote" (an acoustical panel made from recycled newspapers and used here for kitchen cabinets), OSB (oriented strand board, which is a building material made from wood chips and adhesive and used here for flooring), and recycled steel panels and metal (instead of wood) used for stud construction.

The original bungalow had only 650 square feet of interior space.

A view of renovation shows modernist, transparent design.

A solar umbrella stretches over the house, with the second floor patio protected from the sun.

Water ways help regulate rain run-off and provide inviting landscape design.

The interior is light and airy, with an open stairway that helps carry hot air up and out through windows in multiple directions.

8

Forget a Furnace:

Using the Sun, the Earth and Ultra-Insulation

Ted Montgomery, Groundswell Architects

This house is in one of the country's colder climates—rural Vermont—yet it requires no furnace and about one third the energy of similar houses. How is it done? Among the answers to this question are super-insulation, passive and active solar panels, and an earth berm on one side of the house. In fact, architect Ted Montgomery of Groundswell Architects is committed to energy-saving design, and this house and its new addition are good examples of how it can be done. Although this renovation took place some time ago, it has stood the text of time; the super insulated house is, in fact, more and more in demand.

The new addition to what was essentially a 1978 wood salt box used design elements of the original house, but the design of the interior—and the technology—were aimed at creating the perfect thermal envelope.

His combination of insulation solutions includes both active and passive solar panels on the roof and a floor slab of concrete and quarry tile to heat. Hot water is provided by liquid-cooled hot water panels on the south roof, while passive solar energy comes both from roof panels and highly insulated glass facing southwest.

The earth berm, where the building is cut into the hillside, provides natural insulation, since the earth itself is a buffer against the elements. Building into the earth is an old tradition in some cold parts of the nation (remember sod houses?) and, as the earth retains a year-round temperature of 55 degrees, it is a simple way to create warmth in winter and cool in summertime.

But the most effective means is the insulation. The house—though it resembles any number of New England houses—is actually insulated two or three times as much as an average wood frame house today. The heavy insulation traps and retains heat. Though it has slightly larger studs and joists to support the added insulation, it cost little more than the normal cost to build, and the insulation wrapped around every part of it (except the heavily glazed windows) creates an exceptionally tight thermal envelope.

The design of a new addition is open and 25 feet high, so that heat (collected by the masonry floor tiles) rises up through a series of openings in the ceiling. (The only additional heat is generated by a wood stove that uses about 2-1/2 cords of wood each winter—wood that is collected from fallen tree limbs in their own forest.)

The updated home sits on twenty-six acres of pristine—but cold—Vermont landscape.

There are two sets of solar panels on the roof; the higher set is for active energy, the lower is for passive solar energy.

The lower living quarters are set into the earth for ecological and environmental reasons, with the hillside providing natural insulation.

Thermal storage is a function of the masonry tile floor in the kitchen and dining room.

The twenty-five foot high interior in the new addition has balconies and partially open ceilings, permitting heat to rise.

9
A Utah Renovation Creates a Micro-Climate

Angela Dean AIA, AMD Architecture

Architect Angela Dean has recently renovated a house with the goal of creating a "sustainable oasis in the city." Facing south, overlooking a flowing creek, the building sits amid mature trees. The design of the redone house includes nature throughout, creating an opportunity to utilize the micro-climate and "enhance the comfort of the home." From passive solar energy to a ground source heat pump, to vegetation on the roof, this is a prime example of how nature can be central to an eco-renovation.

Among the innovative ideas used by the architect are employing growing nature itself, in the form of gardens on the roof tops, trellises, planters, and balconies. Vegetation helps "to anchor the building to its site" and aids insulation, creating a healthy thermal envelope, as well as natural cooling and ventilation. The design was actually conceived to create spaces for vegetation; the three stories of the house all have small rooftops supporting growing plants on them, with the greenery being both an ecological and aesthetic choice.

Renewable energy was a major aim. There is a ground source heat pump, an energy recovery ventilator (with three-story air flow), and in-floor radiant heating. A close-combustion fireplace (which pulls in air from outdoors rather than from within the house) is an environmentally friendly addition to the heating system.

Recycled and salvaged materials used in the renovation include "Trestlewood" posts and beams, and recycled cellulose insulation. This renovation, using salvaged materials and employing so many natural elements, is a good example of truly sustainable architecture.

This three story house creates a micro-climate with vegetation on rooftops, trellises, balconies, and in planters.

A close-combustion fireplace supplements in-floor radiant heating.

The glassed-in music room has passive solar energy

A series of roofs, awning, and set-back balconies are part of a natural cooling system, through shading and ventilation.

10

Surprising Addition to a Small Urban Home

Shelter Architects

In a post-World War II neighborhood of Minneapolis, this one-and-one-half story home needed expansion without disturbing the streetscape. In an effort to create an alternative to the "sprawling, material and energy intensive larger suburban homes," Shelter Architects devised a plan to add on behind the house with a contemporary and sustainable new wing.

The plan was to open up the original rooms to an ecologically sound tower, providing a significant amount of square footage, as well as passive heating and lighting.

In addition, the original house was updated; it was reinsulated with high efficiency spray foam, and durable low-to-no maintenance finishes were used on the exterior.

The new addition is mostly glass, allowing for natural light and ventilation throughout. An open staircase rises in the glass-enclosed structure. Light is central to this redo: even the new deck behind the addition is designed to be illuminated by the brightly lighted interior. Passive heating and lighting were also created by high efficiency systems, including mechanically operated windows. The library features a passive heating and lighting window bank with a colored internal window well "to color the light and warm the space," according to the architects. The main and upper levels were kept open in case future changes were desired.

A new open kitchen was added. The master bath was fitted with recycled tile, and eco-resin translucent dividers, as well as with fixtures fabricated and installed by local artisans.

This house remodel was one of the first Minnesota Greenstar projects, with an aim of efficient and sustainable use of energy, water, and space.

This compact one-and-one-half story home in an urban area was built after World War Two.

The contemporary addition is in back, sustaining the streetscape and lighting the outdoor area.

The staircase in the new addition is open; it is designed to help provide passive lighting, ventilation, heating, and cooling throughout the house.

The library has passive heating and lighting, and a window bank with colored internal window well "to color the light and warm the space."

The master bath has eco-resin translucent dividers and recycled tile.

11
Thermal Pane Glass Makes a Difference

Mark O'Bryan, Architect

This renovation was designed and built as an addition to a 1960s modernist house. The architect, Mark O'Bryan, was seeking a green solution to saving energy and creating ventilation for the house in the warm summer months. He chose thermal pane glass—lots of it—for its many ecologically sound attributes.

The new addition is "a beautiful glass-enclosed room in a garden setting," says the architect," consisting of a large social dining space, a coatroom, a pantry, and a guest bathroom on the first floor, with an art studio below. In order to be consistent with the design of the existing house, it is characterized by beam, plank, and masonry shear-wall construction, as well as by its emphasis on classic symmetry and pure geometric forms.

The addition has walls that consist of forty-four individually sized pieces of thermal pane glass, which replaced an old carport. Twenty-four-foot long box beams and planking were painstakingly deconstructed and recycled from the carport. They were joined with cedar vertical boards (also reused), and the fourteen-foot high floor-to-ceiling windows. (The architect chose standard sizes to lower the cost of the high performance windows; the total was $8000 including material and installation.)

Of particular note were the advantages of that much glass: passive solar heating in winter, cross-ventilation cooling in fall and spring. A flat seven-foot projecting cantilevered roof provides shade in the summer time.

Other green elements include rain barrels to collect water for the garden and the use of all natural materials (slate flooring, granite countertops), and salvaged materials, like the translucent onyx used for door side lights (recycled from a church renovation).

Also added to the house was a brand new studio space, excavated from below the main level; it provides both a work space and added insulation. The entire renovation cost less the $80,000.

A "before" photo of 1960s modern style house in Kentucky.

in a Kentucky Home

The new entryway shows the relationship between the existing house (behind) and the addition.

Floor-to-ceiling thermal pane glass windows provide light and passive solar heat in winter, and are protected from the sun by seven-foot roof overhang in summer.

An interior view shows salvaged beams, some of the forty-four windows, a double-sided fireplace, and exceptionally high ceiling.

Salvaged wood and translucent onyx light fixtures are among the recycled elements; much of the wood used in the addition was salvaged from the original carport.

12
A Live/Work Retreat Transformed

Arkin Tilt Architects

Photos by *Edward Caldwell*

A rather "nondescript house" in a beautiful coastal California setting was recently renovated into a live/work retreat with the imaginative use of salvaged materials and green techniques. The design, with its varied rooflines, was inspired by California's agrarian compounds.

The architects, Arkin Tilt, created an inner courtyard protected from the coastal winds by the addition of new wings on either side of the house. The outdoor space, which is open to the sun, has a breezeway that can be closed off with a sliding barn door. The renovation also included a solar energy system, adding insulation (polyurethane spray foam), high-performance windows, and a second-story deck.

The use of salvaged materials accentuated what the architects describe as "the ad hoc collection of forms giving each space character." Among the interesting salvaged materials are the exterior siding of old barn wood (given a driftwood stain), and

the reused (and rusted) corrugated metal for outdoor fences and entry gate. The interior floors and ceiling were covered with reused Douglas fir. The sloped ceilings in the newly glassed-in living room were made of fir salvaged from shiplap siding. Other wood was locally harvested.

The kitchen was particularly notable for creative salvaging. The cooktops were restored 1950s Wedgewood (converted to gas), the countertops were made from a saved bowling lane, the sink was recycled, and the knobs on the cupboards were made from stones found on the beach. Another imaginative recycling was the salvaged lintel stone over the wood stove; it was once a limestone fence post. Such reuse is an integral part of the green movement; it adds both character and sustainability to the design.

Varied roof lines. high-performance windows, an exterior sheathed in cedar with a driftwood stain (opposite), and reused rusted corrugated metal on the fence and gate (below) characterize this redone coastal California house.

The lintel stone above wood stove in main entry was once a limestone fence post; the inglenook surrounding it (a favorite design of Frank Lloyd Wright) provides a cozy place to sit.

The kitchen features many recycled materials: counters came from a bowling alley, stove was a restored classic Wedgewood, and cabinet knobs were made from stones found on the beach.

Before and after interior: wrap-around windows and new sloped ceiling made of salvaged shiplap siding give the redone living room its open ambiance.

13

Adaptive Reuse:

New Green Update for a Former Warehouse

Ann Temple ALA, SMNGA Architects

For a downtown setting nothing says "green" more than adaptive reuse, as can be seen in this award-winning conversion. Architect Ann Temple of SMNGA in Chicago has redesigned an old downtown warehouse and elegantly converted its 9000 square feet into six housing units, with a fifteen-foot wide courtyard on its southern façade. The combination of industrial materials, contemporary design, and native plantings make this project an excellent example of what can be done with old buildings—including those that are large and derelict.

Such recycling is, of course, one of the major ways of conserving the past and making aging downtowns sustainable. When such building recycling is combined with imaginative, new, eco-design, it becomes an example for cities across the country that are seeking to upgrade and conserve. Salvage of building elements not only meets the aims of eco-sensitive architects, builders and home renovators; it also offers aesthetic pleasures—particularly to those who enjoy the antique or unusual.

In this building, the choice of metal materials is particularly noteworthy. Many different industrial materials are combined in this project, including steel bar joists (both in the interior and exterior), metal siding, and steel grating. The living units have two-story glass fronts and poured concrete floors with radiant heat, as well as glass block, permeable paving, and other green features.

The design includes a new courtyard with native plantings surrounded by a variety of industrial grills and gratings; the combination is both modern-looking and ecologically sound. The courtyard is an excellent prototype for "deep" structures that do not have street frontage. The redo of this building received a Design Excellence Award.

Industrial materials, including steel bar joists, metal siding, and steel grating, combine with plantings in a sustainable and aesthetic design.

Opposite:
The interior's contemporary design includes overhead steel, glass block, double-story windows, and poured concrete floors with radiant heat, while the open, loft-like upstairs continues the industrial motif with metal rails.

A converted warehouse now has six housing units around a fifteen foot-wide courtyard with steel joists overhead.

14

Light and Airy:

A California Makeover with A Pyramidal Skylight

Obie G. Bowman, Architect

Photos by Tom Rider

Natural light, with its central importance to a home, played an essential role in this home makeover. With a combination of green strategies, the architect, Obie G. Bowman, reconfigured the house, redefined the entry, and freed up interior spaces. The use of natural light instead of electricity was a major objective, as was the elimination of air conditioning.

Working with a 1977 ranch house that had an awkwardly divided interior and a "hodgepodge of window shapes," he devised a makeover using a variety of remodeling techniques. The conversion began with demolishing certain elements. An oversized fireplace that divided the space into "two dysfunctional halves" was removed, as were a narrow central hallway and a number of interior walls.

A new fourteen-foot free-standing kitchen was rotated forty-five degrees and given a skylight, as well as being opened on three sides with glass. The skylight—arguably the signature element of the renovation—is a seven-foot pyramidal form that is intercepted by truncated, perforated metal screens to diffuse some of the bright light. The screens also deflect some of the light to adjoining spaces. The objective was to achieve daytime illumination almost entirely without electricity.

The house avoids air conditioning by the addition of four passive ventilation chimneys that exhaust hot air from the kitchen. Aluminum framed, single glazed windows were replaced with new ones with wood frames and dual glazing.

On the exterior, the house was given a new façade by means of a freestanding redwood screen wall with "openings strategically located to provide privacy, frame views, and create a well-defined entry." (The redwood, like the Douglas fir used, was a local building material.) In addition, the existing cross gables were extended as overhanging trellises in order to shade interior spaces. Light, shade, and ventilation can all be regulated by natural means, without electric power. The overall plan actually reduced the square footage of the home, while providing more open space within.

Before remodeling, this 1977 ranch house had a tile roof and cultured stone veneer.

The front view, after remodeling, shows a redwood screen wall, trellises, and passive ventilation chimneys.

The overhanging gable trellis helps control interior light and shade.

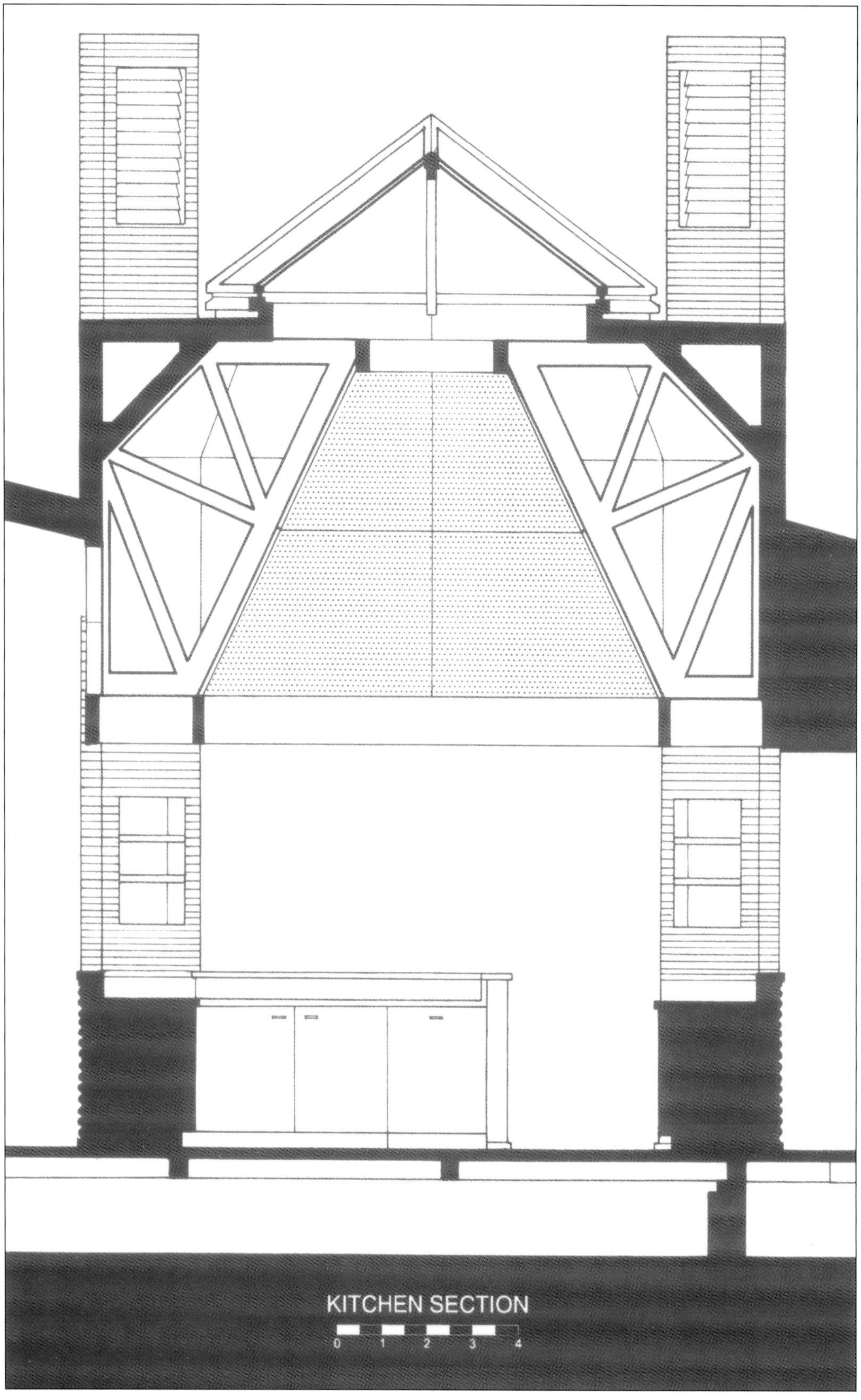

A diagram of kitchen with skylight and photograph of kitchen redo (opposite, top) indicates expanded height and use of natural light.

The skylight, looking straight up through perforated metal screens that help diffuse the light and deflect it to other areas of the interior.

15
The Eco-Restoration of a Prairie Frame House

Locus Architects

This simple frame house on a midwestern prairie recently received an unusual green makeover by Locus Architects of Minneapolis. Using passive solar heating, durable steel siding and roofing, and continuous ThermaCal insulation, the redone house can weather dramatic climate changes.

An addition with a south facing roof now holds sixteen solar panels with effective fixtures providing solar hot water heat and domestic hot water. In addition, a wind turbine helps produce energy (and evokes the image of the old-fashioned prairie windmill).

In the interior, the renovation included salvaging of timber and exposed wooden flooring, with millwork and cabinetry done by local artisans. A final addition was a new stone terrace and the restoration of a fish habitat.

In its rural setting this house looks like a traditional prairie house, but its update makes it thoroughly energy efficient and sustainable. The addition and renovation cost about $500,000.

This picture, taken before renovation, shows a traditional frame house on the prairie.

The "after" photo shows solar panels on the roof and a reworked design of the building, including new stone patio.

Solar machinery provides hot water heat and domestic hot water.

Not far from the prairie house, a wind turbine helps produce energy.

The architects chose durable steel siding and roofing for sustainable exterior.

16
A Green and Passive Addition:
The Butterfly Porch

Rick Harlan Schneider AIA, Inscape Studio

Photos by Dan Redmond

A Maryland suburban house has an original new addition designed to maximize sunlight in the interior, create a usable outdoor space, and provide rainwater run-off outdoors. Architect Rick Harlan Schneider designed a "butterfly porch" with a striking canopy to create a green and passive addition to the home.

This unusual addition is a stand-alone, modern looking porch, which—unlike traditional porches—"pulls away from the house," so that the deck/cook/out area provides space without its roof blocking sunlight into the adjacent kitchen. The butterfly porch has carefully positioned slats and planks, framing views and providing privacy from neighboring properties. The project was built of non-toxic, pressure-treated lumber and sustainable harvested cedar.

Above it is the signature butterfly roof canopy (made of lightweight fabric) which "floats above the crafted box," according to the architect. The canopy is supported by a light metal frame and four posts that are separate from the porch. The four posts were carefully sited to avoid tree roots. At night the orange canopy is lighted from below, making it resemble a hovering butterfly. The space between the house and the porch functions as a deck; the porch itself has screened walls and ceiling beneath the canopy.

The canopy addition protects the porch from the elements; it also redirects water into a planned rain garden, which will redirect storm run-off, as well as aesthetically implementing the design. One technique used for this relocation of water run-off is the rain chain—an idea borrowed years ago from Japan—which draws water from the roof downwards into the garden.

The fabric butterfly canopy sits above the wooden porch addition.

A side view of the addition shows both the canopy and the rain chain which draw water down into the garden.

The porch, with slats of different widths, creates privacy and frames outdoor views; there are both indoor and outdoor spaces with a serving counter between them.

17
A Sustainable, Metal-Clad Addition in a Familiar Shape

Webber + Studio, Inc. Architects

Photos by Tom McConnell

Working with a small, postwar cottage, the architects at Webber + Studio designed a remodel and addition with a focus on neighborhood integration and sustainable practices. The little Texas cottage had two bedrooms and one bath; the architectural plan called for an enlarged living room, a new kitchen, dining room, laundry room, master suite, and nursery. The addition (and a covered porch) adds one thousand square feet to the 870 square feet of the original cottage. The addition is barely visible from the street, maintaining the traditional street-side design.

The architects wanted to tie the new addition visually to the original cottage shape, although the materials would certainly be different with the new section wrapped in a metal exterior. (None of the exteriors have a painted surface.) The metal cladding covers the roof and unfurls over the side, while similar railings of metal accent both old and new parts. The metal roof is held away from the interior with polymer sheets creating an air gap. Below the new master suite is a covered porch, and between old and new sections is a small entryway linking them.

The interior is newly lighted in the renovated section, while the addition has a wall of heavily glazed windows. Non-toxic, no-VOC paints were used, as was a water-based sealer. The design "achieved the owners' desires with practicality and artistry" with this sustainable and unusual design.

This "before" picture shows a postwar, two bedroom Texas cottage.

"After" renovation, the street-side of the house maintains traditional design.

The new addition has metal cladding, a wrap-around roof, and heavily glazed windows.

The metal-wrapped addition encloses new master bedroom, and echoes the shape of the original cottage.

A small entryway ties together original house and new addition.

A light-filled interior ties indoors to outdoors with heavily glazed windows and covered porch.

18
An Environmentally Responsible Makeover of a

Christina Griffin, Architect

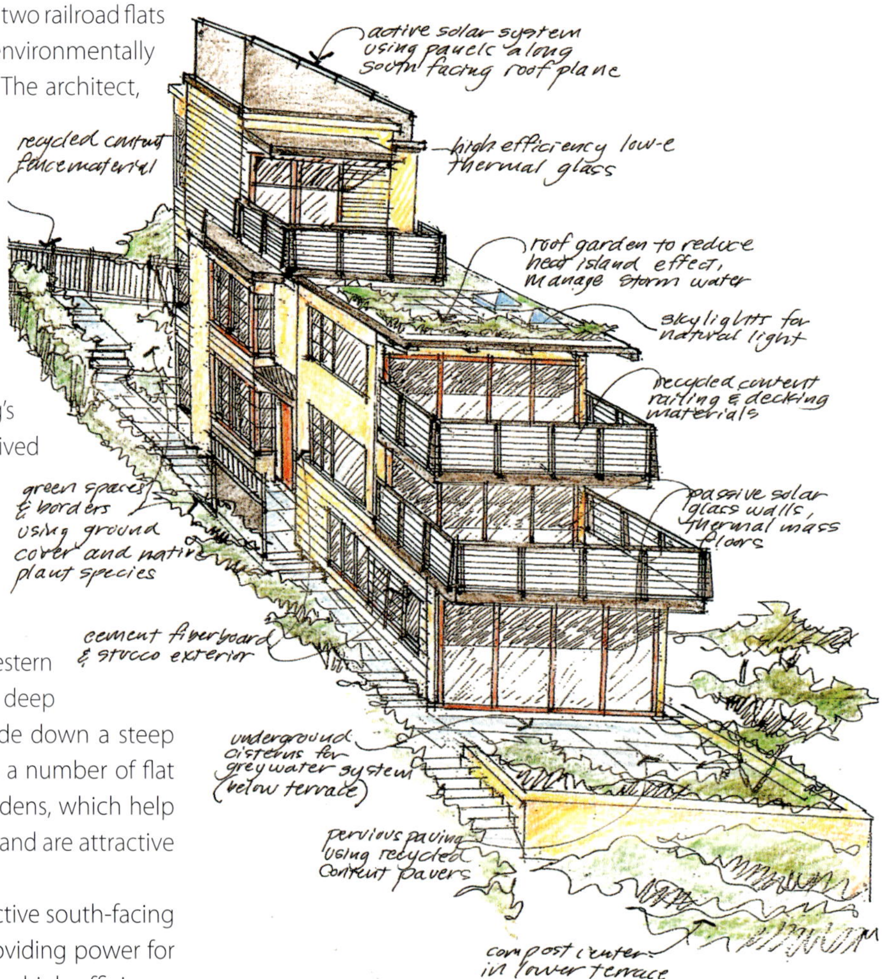

This diagram of River Town House identifies green elements.

A worn, two-family house built in 1910 with two railroad flats atop one another has become an example of environmentally friendly design, now called River Town House. The architect, Christina Griffin, hopes it will be a precedent for reuse in the Hudson River town where there are many such downtown houses overlooking the exquisite river view. River Town House now consists of two energy-efficient, three-bedroom condominiums.

Using a series of energy-saving and economically advantageous ideas, the renovation is worth a close study. The building's rebirth as a green and chic place to live has received an Energy Star rating and a LEED for homes certification.

The redesign made the building into four levels, with a series of high performance, argon-filled glass windows with energy efficient glazing, providing passive solar gain along the western wall, as well as expansive river views. There are deep roof overhangs; a series of tiered levels "cascade down a steep slope towards the river." The redesign includes a number of flat roofs and terraces to accommodate green gardens, which help to insulate, provide storm water management, and are attractive roofscapes for the neighbors nearby.

Solar panels were installed on a heat-reflective south-facing roof and the southern side of the building, providing power for heat and hot water. In addition to solar energy, high efficiency heat/air conditioning pumps are located in different zones of the house to supplement where needed.

Materials include many recycled elements: kitchen tile made from recycled glass, decking made from recycled plastic, salvaged stone and wood framing, and granite for the new patio that was recycled from the old foundation. Even aluminum and copper piping was saved. All interior finishes are non-toxic. In the interior, woods come from sustainable forests.

Saving water is another specific aim of the redesign; the property and the various roof decks are planted with perennials and drought-resistant plants. A rain-barrel, pervious materials for pathways (allowing rain to sink beneath them into the soil), and storm drains that lead to underground water-storage chambers are among the system's water-saving elements.

Hudson River Townhouse

Before and after renovation: the original building dates to about 1910 and housed two families in railroad flats. The updated town house is completely redesigned for sustainable, eco-friendly living.

The interior has windows looking out at Hudson River view, certified sustainable woods, and tiles made from recycled glass.

19
An Architect's House Gets a "Solar Sink:"
A Case Study

David Hertz FAIA Architects Inc.

Architect David Hertz's own contemporary house has been renovated with innovative technology to make it as sustainable as possible. Despite its large size (4400 square feet), the plan called for an environmentally responsible—as well as aesthetically pleasing—redoing of the original house. The remodel was planned to incorporate the latest in solar technology and new materials.

The original house was enlarged over the years; after the renovation it consists of a compound of four discrete two-story buildings linked by enclosed bridges that face onto a courtyard. (The architect describes its style as "Balinese Modern.") Its large size encouraged the architect to search for sustainable solutions.

Using the house as a case study for green design, Hertz decided to use solar energy for most of the house's electrical needs; some twenty solar collector panels were placed on the roof. They help generate about seventy percent of the home's electricity. Other sections of the roof have flat-plate collectors that provide hot water to the water heater, which in turn sends it to a radiant heating system in the concrete floors.

These are not just ordinary concrete floors; the choice was to eliminate carpets ("progenitors of mold and dust"), or wood floor finishes. Instead, a recently developed product called Syndecrete, which contains about forty-one percent recycled content, and is twice as light (but has twice the compressive strength as normal concrete) was used. The Syndecrete floor acts as a kind of "solar sink for passive solar energy transfer," according to the architect. The floor stores up the sun's warmth in daylight hours, so the house does not overheat, and then slowly releases the heat at night. Concrete floors of this type require little maintenance and are environmentally sensitive.

The contemporary design of the compound requires a great deal of wood for the many areas of open rafters, decks, and bridges from one section to another. Recycled and certified sustainable woods used in the renovation included ipe, mahogany and fir.

This building was chosen for a "Sustainable Living Award" by the Eco Home Network.

A complex of four separate, but connected, contemporary buildings has an array of solar panels on the roof.

Balconies overlook an interior courtyard with a reflecting pool in the center.

Le Corbusier
Le Corbusier
Le Corbusier
Le Corbusier
1910-29
1929-34
1934-38
1946-52
1952-57
1957-65
Aluminum

Concrete floors are made of Syndecrete, which functions as a heating agent for solar energy.

Interior views show a large amount of sustainable woods used in rafters, balconies, decks, and enclosed bridges.

20
A Southern Lakeside Cabin with an

Taylor Dawson & Associates, Architect

Photos by Robert Fouts

An eighteen hundred square foot family cabin in Alabama was recently reworked by the architect Taylor Dawson III. Perched on a finger of land surrounded by water, the cabin was dramatically enlarged; the award-winning new addition is an example of how natural light and lakeside breezes can be brought into the interior, using a series of sustainable methods.

The entire house is made of Southern yellow pine, a local, renewable building product. The addition was placed on the front of the existing cabin and separated from the original structure with a familiar Southern design: a dogtrot—an open-at-both-ends, roofed corridor, which provides a constant flow of air and shelter. Three hallways open to the outdoors. The dogtrot is generally darker and cooler than the rest of the house, providing a good workspace and sitting area on a hot day. An open cupola at the top of the house also draws air into the interior.

The major innovation, however, was the design of flip-up windows to regulate air flow into the house. Operated by a simple pulley system, thirty-four of these flip-up windows can easily go up or down, depending on the weather. A separate sash was built for each screened window; they are hinged at the top and a trellis supports them when they are in an "up" position. Each window has its own pulley.

This lakeside cabin, called Land's End, received a one thousand square foot addition. Note the open cupola for ventilation at the top.

Award-Winning Addition

The new living room is mostly glass, with the windows looking out on the expansive deck facing the lake. There are some fifty different sized windows. "If the wind blows from the west," says the architect, "we then lower the windows on the west side and raise them on the east. The low-pressure zone on the east will draw the air through the smaller openings on the west and give us air circulation inside." There is little need for air conditioning, despite Alabama's hot summers.

This design has won a series of awards and has been featured on television. For the family who enjoys the lake front property, it is a welcoming, and almost always cool and green getaway.

The addition was placed at the front of the original house; a dogtrot (a roofed, open-ended corridor) separates the old and the new.

Innovative flip-up windows regulate ventilation and provide shelter from the rain.

A deck, for lounging, and the living room face a scenic lake; natural light and breezes are brought into the interior by flip-up, multi-windowed walls.

21
Building Up, Not Out in the Catskill Mountains

**Andrew Tesoro,
Tesoro Architects**

Set in a particularly scenic mountain landscape, this small house needed to be enlarged. The new design, by architect Andrew Tesoro aimed at adding more space in the interior, while relating the structure more directly to its spectacular mountain surroundings. He described the house as "poorly sited" in relation to the view, and "looking inorganic and out of sync like a glorified trailer plunked down by helicopter."

The architect set about to reorient the house and to enlarge it, using every possible green solution—all within a definite budget. Among the problems to be solved were regular flooding from rain run-off, a poorly positioned set-up for solar concerns, windows facing away from the view of Mount Mohonk, the need for more interior space, and more energy-efficiency.

The decision to build up, rather than to increase the "footprint" of the house was the first green choice. Saving the existing foundation and preserving the shell of the house saved the cost and effort of demolishing and the loss of materials. Building upwards also kept the structure volumetrically compact, and exposed less of the perimeter to the elements (thus preserving more energy). The remodeled house now has two-story glass expanses facing the mountains, as well as an entirely new, elegant appearance.

Other major decisions involved heating and electricity and the possibility of solar energy. Since the site did not lend itself to solar panels or collectors, a different green solution was found for heating and cooling. It included ultra high efficiency mechanical equipment, radiant floor heat, high-tech insulation (Icynene spray foam—a water-based environmentally friendly cellulose to seal the house), and a secondary heat recovery ventilation system. New windows of low e Argon glass were installed, with trellises and fin walls to shield the house from the sun in warm weather. The roof topped off the green renovation; it is made of a standing seam reflective metal to retain heat in winter, and lessen it in summer. The water run-off problem was solved by relocating the garage and sealing the basement.

The final bill for the renovation was about $400,000, of which 15% is directly attributable to green innovations. (The owners expect to easily recoup that amount via tax rebates and lower energy costs.)

A "before" picture shows the original house looking, as the architect says, "like a glorified trailer plunked down by helicopter."

The completed renovation literally raised the roof, "building up, not out."

The exterior is covered by cedar shingles, while the roof is made of standing seam, reflective metal for tighter insulation.

Light is central to the remodel: Natural light and fluorescent bulbs make the interior bright. Huge low-E Argon glass windows face northwest toward mountain view, while trellises and fin walls shield the windows from afternoon sun.

22
The Green Winterizing of a Maine Coastal "Cottage"

Robert Knight, Knight Associates. Architects

This spectacularly set "cottage" on Maine's rocky coast was due for an update to make it able to withstand the harsh seaside weather for year-round use, as well as to redo its infrastructure. While zoning restrictions prohibited new homes on the shore, renovations were allowed, and architect Robert Knight undertook the job on this pre-1920 "vernacular" shingled house.

Though he described it as generally impractical to redo such a building, it was too lovely a house to lose, and he decided to take it down to its original framing. He enlarged it "modestly" and then totally rebuilt it "board by board." Small new additions included a mud room and a new entryway.

Major improvements involved the foundation. Exposed pipes and lack of insulation made the house impossible to heat. The new reworking replaced a rocky crawl space with insulation and concrete, and with PEX tubing for in-floor radiant heat. (The site was not conducive to solar energy.)

Insulating the building from seaside weather was a major factor in this harsh climate. Spray foam insulation was used in the walls and roof. Old windows were replaced with weatherproof, low E glass, and a new airtight framing. A "super-efficient" gas boiler was installed.

In the interior dark beams were replaced with lighter wood and white, insulated ceilings took the place of the original dark-toned surfaces, bringing a new sense of light and air into the rooms.

With this remodeling, the house now can be used year-round; and as the architect says "the building will live another 100 years."

A pre-1929, coastal Maine "cottage" has had a green renovation to make it a year-round house.

The house was dismantled down to its original framing and rebuilt board by board.

Before and after pictures of the foundation show exposed pipes and lack of insulation in the rocky crawl space under the house, and redone concrete foundation with PEX tubing for in-floor radiant heat.

Old windows were replaced with new, air-tight frames and low E glass.

As seen in these before and after photos, the interior was redone with lighter toned wood trim and light walls; the house is now a comfortable year-round residence.

23

An Eco-Redo in New Mexico with a Greenhouse Hallway

Edge Architects

Photos by John Robledo

A collection of sustainable improvements turned this flat-roofed, drafty—and cramped-- house into an energy-efficient, passive solar home with newly visible views of Taos Mountain. By adding to and opening up the original space, Edge Architects were able to create a comfortably sized home. And by using the latest in some new green materials, as well as recycling others, they cut energy bills and brought light and air into the interior.

The addition included a new entrance, a new porch, and a new guest suite. Since the house had a southern exposure, the architects used an innovative design: a glass-sided greenhouse hallway to produce passive solar heating and to improve circulation. This new part of the house is both an energy saver and an aesthetically pleasing addition.

A new roof was a major improvement. The flat and poorly insulated, leaky roof was replaced by a pitched metal roof with blown cellulose insulation and new timber trusses to add height and width to the interior. Exterior walls were wrapped with rigid insulation and covered with new stucco.

Salvaged and recycled elements can be seen throughout the house. From recycled lumber to reclaimed shutters to recycled corrugated metal for the kitchen cabinet doors, reuse was central

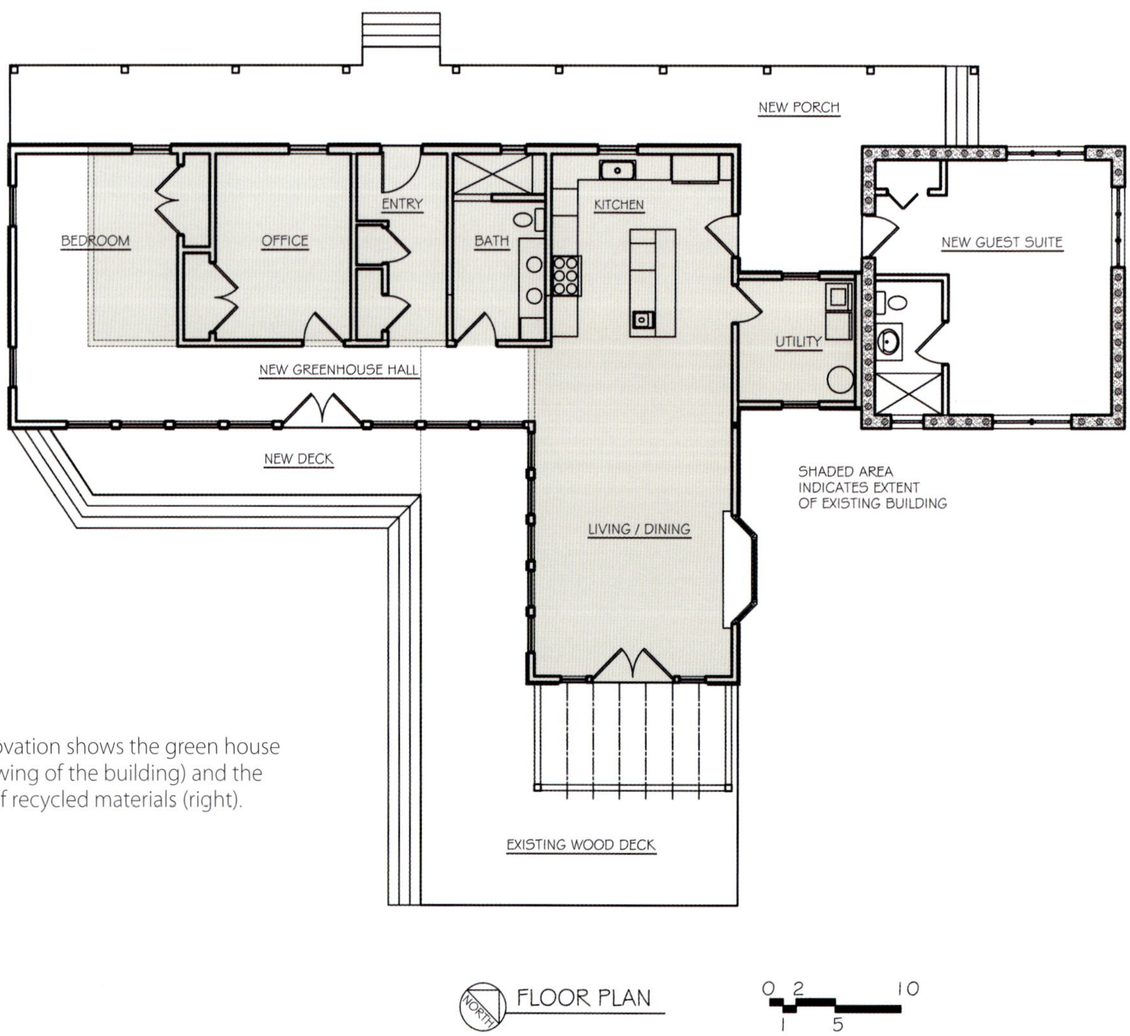

This diagram of the renovation shows the green house hallway (below the left wing of the building) and the new guest suite made of recycled materials (right).

to the remodeling. There are even bits of wine bottles inlaid into cement countertops! The new guest suite was built of recycled Rastia block (an insulating concrete block that uses Styrofoam) and local timber. Floors were made of sustainable bamboo. Many new windows (some with grass-woven shades) opened up the interior to the spectacular mountain view. Paints and plasters were made with natural vegetable dyes.

This ecologically sound renovation is an unusual example of innovative design, as well as imaginative and sustainable reuse.

Before and after: the original house was flat-topped and cramped. Renovation included a pitched, insulated metal roof and additional space created with the latest in eco-materials.

The greenhouse hallway increases passive solar heating and ventilation, as well as adding an attractive feature to the house.

The kitchen uses salvaged materials, including recycled corrugated metal for cabinets, reclaimed bar stools, and concrete countertops inlaid with old wine bottle tops.

24

A Green Redo for a Rural Lifestyle

Leger Wanaselja Architecture

Photos by Kurt Wanaselja and Linda Svendsen

The beauty of nature is important to this house; not only is its site on the edge of the oak forests of Las Trampas Wilderness in California, but two large oak trees in particular have made an important impact on the renovation.

The project was to transform "a run-down, energy-guzzling 1970s ranch house into a showcase of timeless ecological design," according to the architects at Leger Wanaselja Architecture. From deteriorating plywood siding and deck, to a poor energy envelope, cracked vinyl, expensive electric heat, single-paned windows, to minimal thermal mass—this house definitely needed an overhaul.

During the design phase two oak trees fell on the property. They became both a design element, and a great source for ecological renovation. The owners milled them on the site, dried them in their garage (for two years) and then incorporated them into the reworked design. Almost all the finished wood in the house (siding, cabinetry, flooring and fencing) was made from the 6000 board feet of these two trees. Some large branches even became part of the décor.

Energy conservation was a major aim for this "energy hog," Although the original footprint of the house was retained, the interior was remodeled to improve the thermal envelope. New additions included slate, stone and concrete, and extra thick sheetrock (to retain heat in winter and to temper daytime energy swings). Blown-in cellulose insulation on the walls and floors, and rigid insulation in the ceilings also added thermal mass. The addition of photovoltaic panels and a solar hot water system helped lower energy costs. The living room features a recirculating heat fireplace. New argon-filled windows replaced the originals; these "tall windows increase the night-time stack effect venting. Now the remodeled house's energy performance has improved nine fold, and it is almost completely heated by passive solar energy.

New kitchen cabinetry and storage areas are designed for home-grown foods to be canned, dehydrated, or stored. This project is an example of what can be done to transform a rural house into a showcase for "green living"—from using one's own fallen trees to growing food on site.

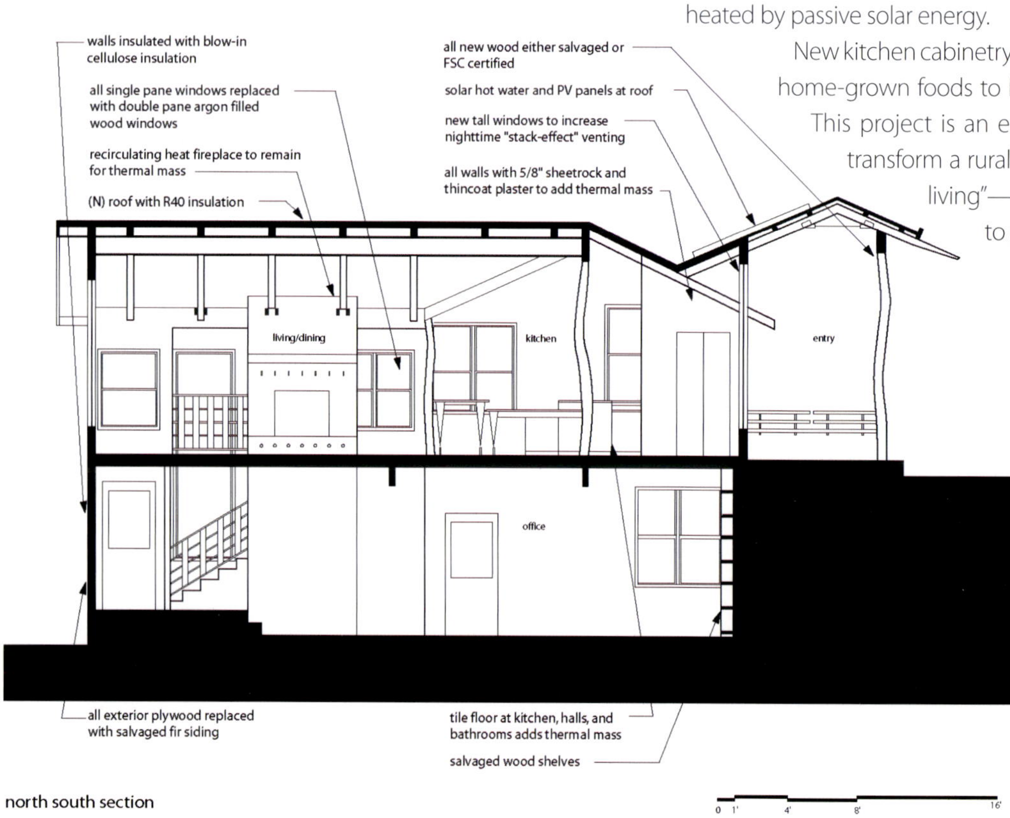

north south section

The floorplan for this reworked ranch house shows some of the elements of the redesign.

The house sits at the edge of the oak forests of a California wilderness.

The roof has solar and photovoltaic panels for energy saving.

A recirculating heat fireplace helps retain thermal mass in the interior.

Two fallen oak trees from the property were recycled and used in the redesign for cabinetry, flooring, table top, and fencing, as well as for aesthetic effect.

25
The Updating of an Early Solar House

**Ted Montgomery,
Groundswell Architects**

Ahead of its time—this unusual and ecologically progressive house was built in 1978, before many architects were thinking of sustainable energy. Here, in the very cold northern Vermont winters, however, practically no additional heat is needed (nor cooling in summertime), due to the use of an original and compelling design and to the latest in engineering innovations.

Set on twenty-five scenic acres, the original house, designed by Ted Montgomery of Groundswell Architects, featured an extraordinary curved roof. It faces south for maximum solar gain. The roof was curved to encourage southern breezes to flow through it, while cold, northern winds blow over it in winter.

The recent renovation, by the same architect, provided more space (two studios, an improved master bedroom, and a deck), as well as several additional sustainable innovations.

A major effort in the renovation was to redo the energy package. Although the house was thoroughly insulated from the start, the use of an earth berm (a hillside built up against most of the first floor wall on the north side) created an even tighter thermal envelope. Passive solar heating is provided by 137 square feet of south facing double-glazing in the roof, while hot water carries the heat through the house in large pipes or tubes. The kitchen, for example, has water tubes on each end of the counter, and there are pairs of larger tubes in other parts of the house, including the music studio. Six floor-to-ceiling water pipes in the living room catch the sun's heat in the day and radiate it at night; they carry stored hot water. Skylights are covered at night with quilted shades. The house is built on a slab of concrete with Styrofoam under it; it too retains heat. Additional changes to the house include replacing the original roof with a curved metal material, exchanging the exterior siding for cedar shingles, and providing a new insulating coating to the first floor.

These cutting-edge technological innovations have not diminished the design or ambiance of this truly remarkable looking Vermont country house.

The original 1978 contemporary-style solar house in snowy Vermont had a distinctive design. Cold wind blew right over the curved roof.

Shown in a lush summer landscape and recently updated by the same architect, the house has new siding, roofing, and an addition, as well as many technical innovations

The east elevation shows an added earth berm alongside the house and ventilating skylights in the roof.

A tower facing south-east (accommodating art and music studios) has ample sunlight.

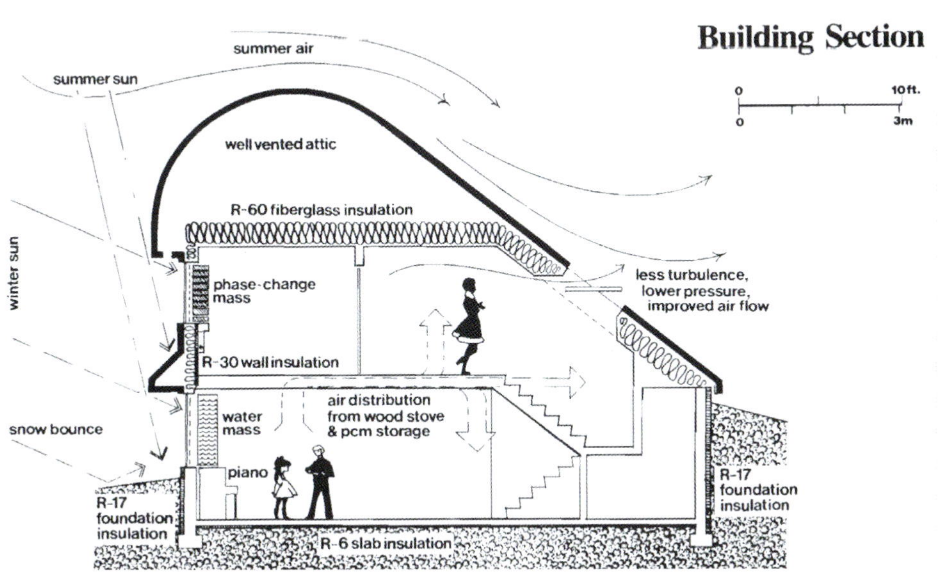

This diagram shows how air circulates through and above the house, and how the design takes the climate and weather into consideration.

26

Low-Impact and Sustainable:

An Old Warehouse Redone

Lake Flatow Architects

Photos by Paul Hester, Hester and Hardaway Photography

The architects at Lake Flatow believe both in sustainability and low impact architecture. Needless to say, the salvaging of an old building is the ultimate ecological choice. This adaptive reuse of an old warehouse is an example. Built on the edge of downtown San Antonio in the 1920s, the warehouse was saved from the wrecker's ball, as well as from a disastrous fire. The outcome of careful redesign and imaginative reuse can be seen in the extended live/work loft it is today.

Working with the fire-damaged remains of the warehouse, the architects decided to make their renovation as simple as possible, with minimal environmental impact, low energy use, and recycled materials. They replaced the damaged roof with a "saw-tooth" roof (much like those on nearby industrial plants), which provides natural light in abundance. They created a steel support structure that is both practical and aesthetically pleasing in the large open interior. A mezzanine added to the loft's design. The center of the building has become an interior, enclosed courtyard, with a pool and greenery surrounded by a ten-foot wall. The design even called for the salvaging of fire-damaged, rusted steel walls, which contrast with nearby smooth plaster and add to the ambiance and reference to its industrial past.

Glass is central to the design; some recycled, some bearing small ceramic spots that reduce heat gain and glare, some made of milk glass. Large glass panels sit in bays, flooding the interior with light. The new saw-tooth roof is fitted with alternating clear and darkened glass. The owners rarely need air conditioning or electric light during the day.

The floor is an unusual combination of tobacco-colored cement mixed with used crank case oil, giving a deep, warm-toned impression that also reflects the natural light from the many windows.

The loft (which encompasses 12,000 square feet) serves as both a home and an office, with elements of its warehouse use still visible in ingenious parts of the design. It suggests the value of conserving the past –particularly in America's downtowns—and of making the adaptive reuse sustainable. But this is not just an exercise in reuse and eco-remodeling; the finished loft is an example of elegance and fine design.

Before renovation: a one-story 1920s industrial warehouse in San Antonio was saved from fire and demolition.

The chic, redone building is now a large, open-plan loft with a mezzanine and an enclosed courtyard.

Before and after scenes of the interior show a simple but elegant remodel, filled with natural light.

The unusual, tobacco-colored floor is made by combining cement with used crank case oil.

Sustainable features include a new peaked roof that allows hot air to rise, while eliminating some air conditioning, sand-plaster walls that absorb heat, and overhead windows that bring in natural light.

Sustainable features include a new peaked roof that allows hot air to rise, while eliminating some air conditioning, sand-plaster walls that absorb heat, and overhead windows that bring in natural light.

27
"Latest and Smartest" in a House Update

Angela Dean, AIA,
AMD Architecture

This update of a 1960s house takes advantage of the "latest and smartest" innovations in green building. At the same time, it creates spaces for outdoor living and optimizes views from the interior. The revisions called for a new layout to benefit energy savings, technological advances, the best eco-paints and finishes, and reclaimed materials.

Changes in layout were aimed at passive heating and cooling. Windows to the east and west were reduced, while an open living plan with a raised ceiling, clerestory windows, and operating skylights for passive convective ventilation added to the light and airy atmosphere (and reduced the need for air conditioning). Shade trees were preserved and overhangs also helped with cooling.

Solar energy is central to the design. Roof-mounted solar panels provide heat for hot water and the radiant heat system. (Solar panels are linked to a water coil within the central air handler to supplement heat on cold winter days.) The roof, which carries the solar panels, is a "cool roof"—a white metal membrane surface with a low slope which reduces the heat load.

Thermal barriers were created by advanced insulation and increased wall thickness. A combination of open-cell and closed-cell insulation foam was sprayed on all exterior walls and roofs. Insulation was also added between the floor framing and radiant heat tubes. Within the tightly sealed thermal envelope are a variety of innovative ways to bring fresh air and ventilation into the home. Among them are electrically operated dampers in each zone, a kitchen exhaust hood linked to an operable fresh air intake system, transoms over interior doors to maintain cross ventilation, a multi-speed blower in the basement, and an energy recovery ventilator for the best indoor air quality.

The importance of water conservation and low energy use were not forgotten. The plan calls for recirculating hot water loops, low-flush and low flow water-efficient fixtures, and a drain water heat recovery system, which is used to preheat the fresh water piped to the showers. A low-water garden is also planned.

The finished project had one more green element: whatever materials were not used in the reconstruction will have a future use through a salvage company. And even the leftover bricks will become landscaping pavers. These many innovations—both technological and architectural—suggest the direction eco-renovations will take in the future.

The renovated house has a white metal membrane covering the roof, creating a "cool roof" at a low slope to reduce heat load.

Roof-mounted solar panels provide heat for hot water and the radiant heat flooring system.

A soaring interior with a raised ceiling uses reclaimed timber for posts and beams.

High, operable, clerestory windows and skylights add to passive convective ventilation. The kitchen features an exhaust hood linked to a fresh air intake system to prevent "negative pressure" in this tightly sealed home.

28

Opening Up a One-Story Structure to Light and Space

Claremont Environmental Design Group, Architects

The existing small, 1377 square foot, one-story house had limited access to ventilation and to light, and was partitioned into small spaces. The owners, with a strong interest in music, also wanted a home that would "integrate their passions and interest into their built environment."

The new addition of some 1200 square feet almost doubled the space of the house. It was designed to add three bedrooms and a bath above, and an enlarged living room (capable of hosting performances). The imaginative solution was not only to add windows (double glazed and operable), but to make an inverted glass roof, thus maximizing natural daylight and ventilation. High north and south eaves slope down to meet at the center, bringing daylight into the interior, as well as drawing the occupant's eyes "up and out to the spectacular views."

The placement of the addition on the north and west sides of the building allowed new vaulted ceilings in the downstairs' newly spacious "great room" (for piano, musical performances, and a large collection of books and music). The original office and library were moved to the north side. The whole project, according to the architects, "promotes the lifestyle of the clients by providing a vertical separation of private and public living space."

The design of the finished renovation and addition "borrows elements from the older structures" in the area thus fitting easily into its neighborhood.

The new addition rises above the original small, frame house.

An upstairs addition was placed on the north and west sides of the structure, with an inverted roof, allowing vaulted ceilings in the main living room. The completed house borrows elements from the design of neighborhood homes.

Dual-glazed windows and doors open onto the patio, bringing light and air into the interior.

29
Reclaiming a House in Texas Hill Country

Bercy Chen Studio LP, Architects

What began as a modest renovation turned into a much more elaborate redo, with recycling as a major element. The reuse of natural and man-made resources characterized this changeover from a somewhat ordinary 1970s house into a glamorous, highly sustainable home. The renovation included both interior and exterior work, as well as a dramatic system of rainwater recycling and a redone garden of native Central Texas plantings.

Among the most dramatic changes were the expansive decks with glass railings (so as not to interfere with the hillside views,) and a new front façade made of recycled glass blocks. (Glass block is an increasingly popular sustainable building material, which provides light and insulation, in addition to privacy.) Shading devices were also installed on the south side of the house.

Another interesting technique that "dematerialized the monolithic nature of the original house" was the use of slats installed as rain screens. In fact, the saving of rainwater is both a measure of the house's sustainability, and a notable design element. Harvested rainwater is stored in pools and reservoirs and a series of cascading ponds and a reflective pool serve as part of the collection system. The falling water provides both an appealing and dramatic element to the exterior, and "a peaceful transition between the landscape and the architecture," according to the architects. The harvested water—in a dry climate—cascades over a concrete pool to a cistern below.

Using the garden principle of xeriscaping to minimize the need for water, the gardens around the house were planted with native trees, shrubs, and ground covers "to preserve the characteristic of the site as much as possible and to retain the essence of the landscape" native to the Texas Hill Country.

This picture shows the original 1970s house before the recycling and reconfiguration of the house and garden.

The redone house has south and east elevations, with an extensive water recycling arrangement that functions both as a sound water-saving system in a dry climate, and as an aesthetic element of the exterior design.

Native plantings are used in the garden to maintain typical Hill Country ambiance, and to minimize the use of water.

The new design of recycled glass block on the northwest corner of the house provides a combination of light and insulation, as well as privacy.

An extensive series of decks with glass railings are illuminated at night.

The contemporary interior includes dramatic seating in the living room.

30
Translucent Walls and Other Surprises in a Reworked

Locus Architects

The most surprising thing about this reworked 1950s "rambler" is the spacious, two-story, loft-like interior. It is unexpected, and just part of the dramatic redo, designed, according to the architects, to be a "synthesis of fanciful urban loft and practical family base camp." In fact, this is a showplace of sustainable design, using green technology, recycled materials, and environmentally inventive ideas. It has been viewed by many thousands of visitors who are eager to see its eco-conscious, low-maintenance, architectural design that is "both economical and expressive."

For example, the walls are complex: on the exterior a combination of ventilated, translucent polycarbonate sheathing covers the exterior, providing passive solar heating, translucency and "dramatic transformation of facades from day to night." Materials include recycled billboards discarded by the highway system; (these are highly durable—and a mounting waste problem). The translucent siding allows light to transform the space during day and night—both from outside and from within.

An advanced vapor barrier/insulation system fills the walls, In an area where moisture in the walls and mold are prevalent this system was an important innovation. In the interior recycled chalkboards form part of the design (and can be used for drawing on).

The loft-like ambiance is created by the two-story height, the steel superstructure, and its components: a floating catwalk with a translucent floor and a hanging staircase. Materials throughout are sustainable or recycled: there is cork flooring, reclaimed redwood trim, ipe decking and steel roofing, and reused plumbing fixtures. The architects describe themselves as idealists, and the project as " a rich material collage of thrift."

Known as "Nowhaus," this reworked, sustainable, single-family home in Minneapolis is remade from a 1950s "rambler."

1950s House

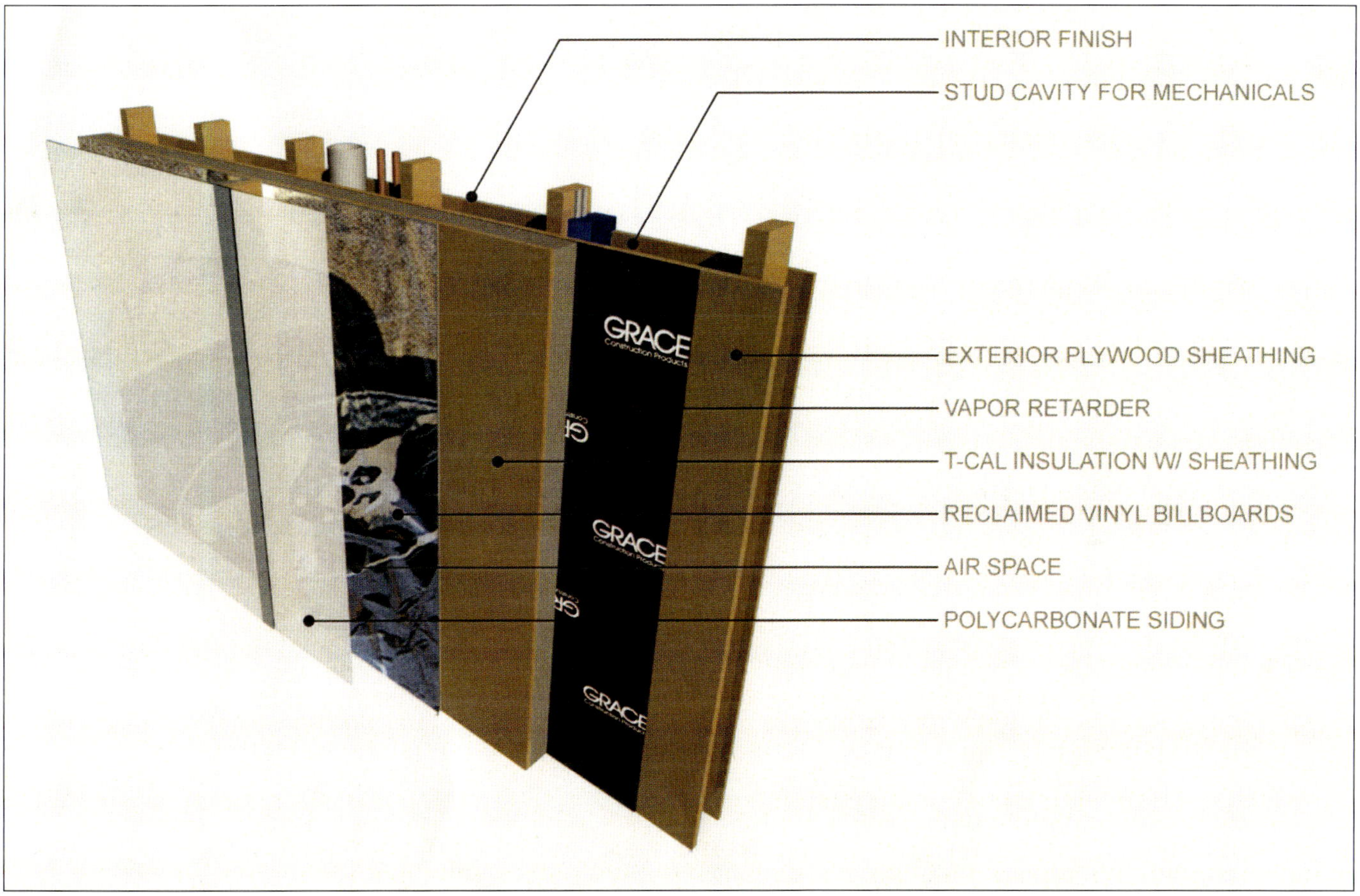

This diagram shows the complex use of materials in the walls, with an advanced vapor-barrier/insulation system within and recycled billboards on the exterior.

The exterior is clad with translucent polycarbonate siding; decking is made of ipe wood.

The architects' aim was to synthesize a practical family home with a "fanciful urban loft" and to celebrate sustainable city living. The surprising interior includes a floating catwalk (below), steel superstructure and "a contemporary sculptural, hanging staircase (opposite, bottom)."

31
An Urban Row House Gets a Makeover

Andrew Tesoro,
Tesoro Architects

Before renovation, the dilapidated 19th century row house near downtown Brooklyn had four stories, but the top floor needed a code-compliant makeover, and the whole building needed conversion for a more sustainable future.

A combination of innovative new additions and salvaged materials make this redo a prime example of what can be done with an antique urban house that is badly in need of updating. "Our primary goal was to reuse and recycle all the materials we could salvage," says architect Andrew Tesoro.

Major renovations took place at the top of the building. The faulty fourth floor was demolished. The brick veneer and heavy timber joists were salvaged and reused for rebuilding the three lower floors. Two new floors were added, as well as a roof deck. The new floors provided space for new bedrooms, two baths, and a yoga room.

The roof terrace was shielded from the sun with a trellis that supported fourteen solar panels, tilted to the back garden of the house. The panels were tilted at the optimal angle—fifteen degrees—for maximum efficiency, with the aim of providing all of the hot water and some electricity. The roof deck, in addition to providing solar energy, also serves as a "luminous and beautiful space" with terrific views of the city and the Verrazano Bridge.

Another problem resolved by the renovation was a lack of interior light. New additions included recessed windows (to shield high sun), low "E Argon" glass, and transom glass above the doors to let natural light into windowless spaces. The reconfiguring of the deep floor plan of this typically narrow brownstone involved consolidating all of the plumbing in the center of the building.

Materials in the renovation included spray cellulose insulation, reconstituted glass countertops, Brazilian ipe wood for decks (ipe is a dense wood that does not splinter, and is considered to be—like bamboo—an ecologically sound choice). Bricks were recycled from the original fourth floor veneer. ("We carefully dismantled the old bricks," says the architect, "and put back all the whole ones. Some bricks broke, which allowed us to enlarge the windows!" The completed renovation maintains the historic streetscape, while providing new, ecologically sound, and aesthetically pleasing additions above and behind the house.

Before renovation, this 19th century Brooklyn brownstone needed a new fourth floor, more interior light, and an energy-saving makeover.

The "after renovation" photograph shows the addition of two new stories, a floating solar panel rack, and a roof deck.

Many large windows, designed to bring light into brownstone's interior, can be seen in this view of the redone garden side of building.

A solar panel rack can be seen at top of the building.

The newly light-filled interior is the result of larger recessed windows made of low-E Argon, and new transom glass over the doorways and staircase. (above and opposite)

32
Green Restoration for Living and Working in an

Ann Temple ALA and Ken Schroeder
ALA, SMNGA

Certainly the most ecologically sound plan within a city is to renovate existing buildings. Architects Ann Temple and Ken Schroeder are experts at taking old industrial buildings and reusing them, while retaining as much as possible of the original materials and atmosphere. This 1930s power station in Chicago is a case in point. Now a live/work site for a writer and a photographer, the 2900 square foot building has ample space for studios and living areas. It has twenty-four foot ceilings, and a mezzanine reached by the original staircase and a catwalk.

The architects retained as much as they could of the power station's architectural elements. They liked the unusual glazed white brick interior walls, the copper-framed windows, and the industrial ceiling grid. They added some room divisions, two fixed glass panel garage doors (which open to a side garden), and new distressed and colored concrete floors equipped with radiant heat. The plan actually reduced the square footage of the floor space, while compensating with double height spaces. New windows were added

The kitchen was created around a 5 x 9 foot stainless steel island; the countertop is a slab of Ubatuba granite. Bathrooms have mosaic flooring made from recycled marble shards obtained from a stone yard. Many other elements are recycled or salvaged.

The renovation can be described as a combination of old and new, and rough and finished. For example, peeling paint on the ceiling was retained, while new electrical and heating systems were added. Smaller shapes throughout the interior are round—including sinks and mirrored portholes in the doors—adding design continuity to the large open spaces. But the overall ambiance of the industrial interior remains, with catwalk, concrete floors, original surfaces, and high ceilings, as well as the giant open space.

Photo taken before renovation shows a 1930s, red brick, former power station in downtown Chicago.

Old Power Station

Aluminum garage doors were added to bring in light and open the interior to adjacent courtyard.

In the interior, an industrial aesthetic was retained with a catwalk, poured concrete floors, copper-framed windows, and white brick glazed walls. The kitchen island is made of stainless steel (opposite, top), with countertops of granite.

33

The Conversion of a 19th Century Barn to a Vacation

Charles W. Moore and Mark Simon, Centerbrook Architects and Planners

Photos by Norman McGrath

Today, salvaging of old but usable buildings is one of the prime elements of the eco-movement. But long before such conversions were popular, Charles W. Moore and Mark Simon of Centerbrook Architects and Planners undertook the renovation of a large but decrepit barn.

Originally built circa 1840-1869 on a Maryland farm near the Choptank River, the barn had distinctive ballast stone walls with hand-hewn wood above, a metal roof, a long shed running along its southern wall, two silos, and a yawning openness within. The owners wanted to use it as a second home for their family of five. They also wanted to retain its character, with the original stone and wood to remain visible from the interior. And the uninsulated barn had to be energy efficient and generally receptive to the sun.

The solution to the renovation of this aging building was an unusual one. They built new walls outside the old frame walls. The new walls, which were attached to ledger strips and bolted into place, were thoroughly insulated, and covered with a new exterior of rough siding, left to weather. Meanwhile, the antique wood, which helped to give the barn its rustic look, remained in the interior. Similarly, the roof rafters were too old and weak to hold up another layer. T-section rafters were added outside the roof, creating space for insulation under a new metal roof.

A major design decision was what to do with the southern wall. The shed running the length of the barn blocked a handsome view of the river and cut off winter sunlight to the interior. The architects chose to strip the siding and most of the roof off the shed, leaving a trellis-like frame, which supports a covering of summer vines, but still provides some shade from the summer sun.

The southern side of the roof became the site for five solar collectors (for hot water), while a conventional oil boiler heats the house. When the house is unoccupied, heat from solar collectors is circulated. Concrete slabs and massive masonry store heat and level temperature changes.

The interior has two living floors and a loft with a catwalk. Three bedrooms, kitchen, and dining room are on the ground floor that opens onto a terrace. A stucco chimney stands the full height of a light well. The strikingly open interior allows spectacular views and sightlines; the original barn siding, rustic timbers, antique stone, and fireplace (featuring metal roofing salvaged from the shed), are constant reminders of the structure's rural origins.

The beautifully set, nineteenth century Maryland barn had a tin roof and two silos before its renovation.

Home With Solar Collectors

The renovated barn has new exterior siding with the original retained in the interior. The southern side of the barn has the trellis-like remains of the shed and solar collectors on the roof.

The dramatic, spacious interior is three stories high, including a loft with a catwalk and exposed beams.

Barn siding, the antique stone chimney, a fireplace made of tin salvaged from the shed roof, and barn doors were all recycled from the original building.

34

A Straw-Bale Addition:

An Inexpensive, Sustainable Building Technique

John Ringel, Architect, Jersey Devil

Photos by Shekinah Rae

One of the most ecologically sustainable and popular techniques now in favor in mostly rural areas is straw bale building. Straw meets the criteria of being local to many non-urban communities, few trees are cut down, and it is a great insulator, sound proofer, and water and energy saver. Above all it is one of the least expensive building materials.

Before renovation, the 150 year-old stone house had an inefficient wooden addition (seen on the right of the house).

New dormer windows were added to the original house, and a new energy-efficient addition was built with straw bales within the walls.

Straw bales—what is left after production of grains-- are an agriculture waste product. They have a high concentration of silica, and do not decompose easily. The bales are set within timber framing and sheathing and installed as infill; the finished wall has a stucco surface on the exterior, and plaster covering the interior. Bales weighing about 60 pounds each are stacked atop one another; each bale averages a foot and a half in height and about 3 feet in width. A stack of six or eight bales will cover the wall area. (It should be noted that straw bales have to be thoroughly dry before use in building, or else they have a propensity to mold and cause structural—and health—problems.)

But many architects are finding straw bale building economical and sustainable; John Ringel of Jersey Devil designed a straw bale addition for the 150 year-old house seen here. He removed an inefficient wooden 20th century addition and replaced it with an energy-efficient and natural structure. The addition included an exposed timber frame (filled with the bales), and covered with stucco on the outside and lime plaster within the house; a living room was added below a new master suite, as well as a one-story kitchen and a stairway and bathrooms. The new stair tower went up to the attic space which had new dormers added, and served as an additional sleeping/play space.

Decorative trees and branches can be seen in the interior. A wide plank floor attests to the antiquity of the original house, while the emphasis on low energy use and sustainability is thoroughly contemporary.

Straw bales, set within timber framing, can be seen in this interior of the walls.

The new living room features lime plaster, structural trees, and a wide plank floor.

The kitchen is rustic in style, but contemporary in its emphasis on energy saving.

35

Small Changes Matter:

Redoing a Brooklyn Townhouse

Ryan Enschede. Architect,
Ryan Enschede Studio

Renovations to a recently purchased townhouse were small in scope, but resulted in a much greener building. The owners needed a lighter, more energy-saving interior, a redone kitchen, a more usable back yard, and a new entrance to the basement and back door. They wanted to retain the style of the existing house, but to work in as green a manner as possible.

The architect, Ryan Enschede, describes this renovation as "an accumulation of relatively small choices." All of the materials used were readily available and met the highest eco-standards. In addition, the use of recycled materials, refitted windows and doors, and new (but recycled) cotton insulation inserted into the wall cavities all contributed to energy saving. New rigid insulation over the outside of the walls (between plywood sheathing and siding) was added as well.

To increase daylight in the interior, the kitchen received new larger windows and both kitchen and shed were given skylights. The dining room now opens onto a very light kitchen (with an outdoor trellis to help with summertime shading). Compact, fluorescent

This Brooklyn townhouse has a small back yard with a new trellis to provide shade in summer, and an ingenious new access hatch to the basement.

Kitchen renovation includes a new skylight, recycled glass countertops, and compact fluorescent lighting.

Below:
New openings between the rooms bring daylight from the newly bright kitchen into the dining room.

lighting fixtures were another addition. The kitchen makeover features reused 2 x 4 lumber framing (salvaged from the original dropped ceiling) and recycled glass countertops.

In the small back yard a new staircase and an ingenious new hatch entrance to the basement were part of the renovation. The final design incorporated an unusual gas-sprung-driven device for the hatch.

The owners and architect were equally concerned with the challenge of waste disposal for renovations like this one. They tried to recycle whatever materials they could (including the aging parquet flooring); other leftover materials were taken to re-use demolition sites. (These are still somewhat rare in some parts of the country.)

This town house make-over points to the importance of small, but incremental, changes that can contribute to a more sustainable and ecologically sound building.

36

"Recycling is the Original Shade of Green"

Jennifer Siegal,
Office of Mobile Design
Photos by David Lauridsen

As someone who has made her name synonymous with the use of "found" materials and unusual ways to employ them, Jennifer Siegal of the Office of Mobile Design is a great spokesperson for sustainability. She uses a variety of unexpected materials in building and renovating. (See page 14 for her use of surplus shipping containers and old grain trailers.)

Most of the things in her house she says are "either found objects or cheap." She finds materials at demolition sites and in dumpsters, or in negotiations with builders and factories. She uses this site to try out new ideas and to text products and imaginative solutions to sustainable architecture.

After buying a modest 1920s California bungalow, she began renovations and expansion ("The place was like a rat maze," she comments). She added to the house with a shiny white, drop-frame aluminum trailer with its original mahogany floors. (She originally purchased the 200 square foot trailer for $1500.) Into this salvaged addition she put large glass doors that she had custom fabricated, and kitchen doors recycled from a grocery store. Other found items included a salvaged Sub-Zero® refrigerator found in a dumpster, inexpensive bamboo flooring, a bookshelf made from scrap lumber, and another extra room that was once a moving truck into which she cut windows.

While all of these changes to the original house were less expensive than buying new materials and appliances, she notes that she has sunk more money into the rebuilding than starting with an entirely new house. "It's about sustainability," she says. "You can start from scratch or you can work with what you have." Her originality and imagination have made what she has both unusual and sustainable.

An aerial view shows the original 1920s bungalow and the recent trailer addition.

An interior view of the 200 square foot moving trailer that was used to enlarge the small house.

A garden view of the kitchen/dining room and patio, shows the double doors recycled from a grocery store.

The addition of a new bathroom connects the original house to the trailer.

37

A Rooftop Pavilion Updates a Washington Townhouse

Rick Harlan Schneider AIA, Inscape Studio, Architect

Photos by Dan Redmond

An original solution to an eco-renovation of an attached townhouse was to build on top of it. The owners of this building wanted to have solar power, and a renovated kitchen, dining room, and bedroom, but they had no budget for elaborate additional design and engineering. The project was to create a roof deck, a kitchen renovation, and an efficient mechanical system for heating, cooling, and conserving water and energy.

The solution by architect Rick Harlan Schneider was an unusual one: a rooftop pavilion that angles out over the street of brick row houses. (Among its assets, is a great view.) The pavilion is made of stucco and it tips lightly to one side to draw and conserve rainwater into rain-barrels—more than enough to water gardens on the rooftop and balconies. The pavilion brings light into the house interior (via stairwells and an operable window wall).

Cooling and heating and air circulation are primary considerations. The pavilion acts as a "solar chimney," with built-in settings that provide cooling and fresh air to the house. It has overhangs to provide shade during the day. It also has a clerestory window and black interior walls (to create a super-heated chamber). Air intake is on the lower, cooler levels through shaded basement window wells, and there are folding operable windows to create air flow throughout the building. A wall of windows in the pavilion makes a beautiful sitting area, while cedar louver doors allow airflow to the solar chimney (while screening it from view).

All natural materials were used in woods and finishes and sealers. The countertops of the redone kitchen were reused. Cork flooring (an easily renewable material) and deep, rich colors characterize the newly done interiors. Other unusual features include an alternative tread stairway (for a tight space) with a closet tucked underneath. In the newly designed kitchen, light fills the space, not only due to French doors and a removed wall, but to glass-front appliances, and—of particular efficacy—back painted glass. Non-polluting features include a low-VOC, waterborne sealer, and ipe flooring coated with natural tung oil. (Appliances were replaced with new energy-efficient models.)

This is an ingenious retrofit, using the latest in technology. Although the architect found few such designs in his research, the combination of solar plan, air intake, and striking design has worked out exceptionally well.

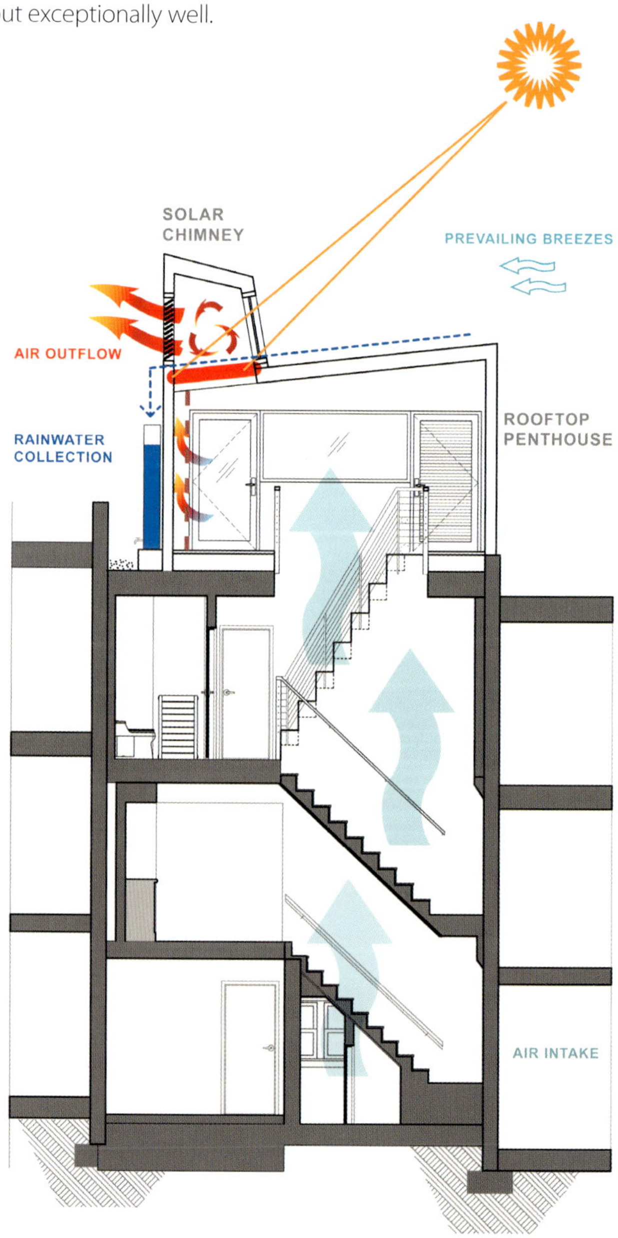

This diagram shows how air and light are circulated in redone building.

The renovation of a townhouse in Washington has a new rooftop pavilion, which provides solar power and tips to one side for rainwater collection.

Previously unused rooftop now provides fresh air and great views; Louver doors are made of cedar finished with a water-borne sealer; siding and decking are certified sustainable.

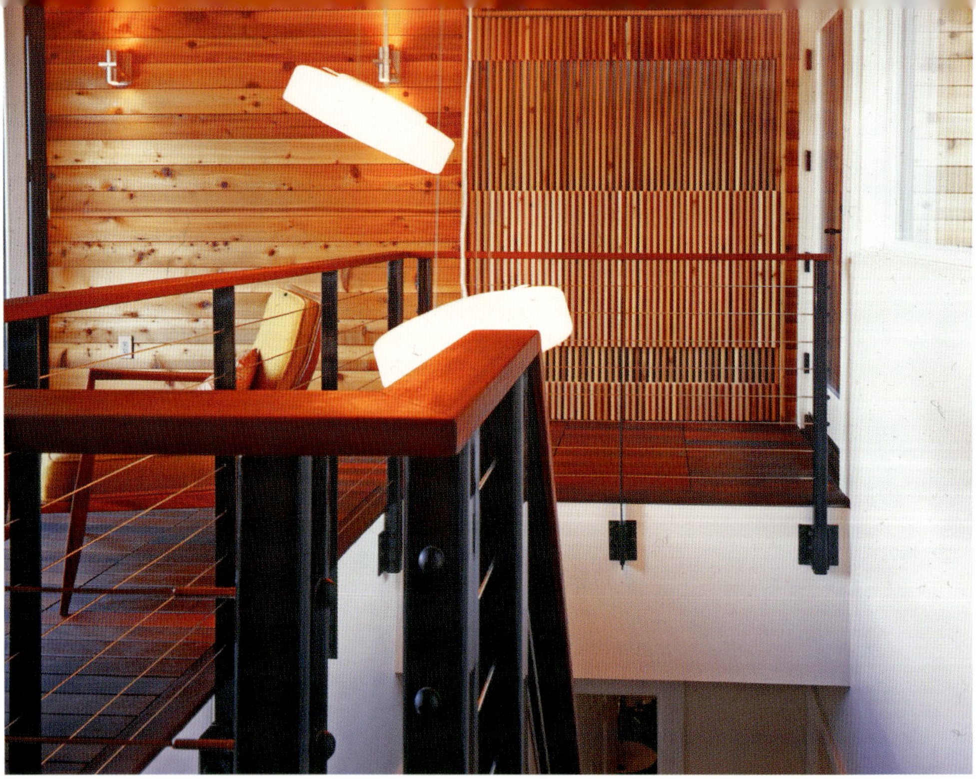

The newly redone kitchen now is now filled with natural light; it has energy-efficient appliances, back-painted glass, and French doors.

An unusual alternating tread stair squeezes into a narrow space, but has a closet tucked under it.

38

Green Choices:

Green Materials Inside and Out in a Long Island Renovation

Elizabeth Thompson, Architect

When architect Elizabeth Thompson undertook the renovation of a small 1950s ranch house on Long Island, she made numerous ecologically sound choices, particularly in her selection of materials. In fact, this modest house suggests a textbook list of green materials.

The existing fiberglass exterior insulation was replaced with "Knauf EcoBatt" insulation, which is made from rapidly renewable organic materials that do not contain formaldehyde, plenol, acrylics, artificial colors or petroleum-based materials. The vinyl siding of the original house was replaced with natural cedar siding and stained, rather than painted. All of the windows were replaced with high performance "tilt-wash" thermo pane, "Low-A Sun" coated high performance glass. A light colored roof was added to absorb less heat and to increase energy savings. Fifteen photo voltaic solar panels (each 2' x 4') are being installed. on the south-facing roof.

Interior doors were all replaced; the new ones are made of solid core pressed wood particle, both for increased sustainability and greater stability in expansion and contraction. New kitchen counters were made of natural ground quartz in a cement binder.

Every room has a new "Altus" ceiling fan for more efficient heating in winter and cooling in summer (thus avoiding the use of air conditioning.) The existing hot water baseboard heating was extended with cast iron, in lieu of aluminum baseboard, for greater efficiency. The previously polyurethaned oak strip flooring was refinished with a water based acrylic.

In choosing these many green options, the architect demonstrates how a simple ranch house can be turned into an example of eco-renovation with a minimum of dramatic changes.

Before: this vinyl-sided 1950s frame ranch house has a painted exterior and small windows.

After: A stained, cedar-sided, and updated exterior has new, larger, high performance windows and redone entrances.

Solar panels on the roof, ceiling fans in every room, and cast iron baseboard radiator are part of the new heating and cooling program.

High performance, tilt-wash, thermopane windows replaced the smaller original windows.

39

Sun and Air:

The Reworking of a Sonoma Coast House

Obie G. Bowman, Architect

Photos by Tom Rider

The magnificently set house on California's coastline was built in 1971 and had been "haphazardly remodeled" several times, leaving a warren of rooms, different levels, and multiple stairways. The renovation was aimed at opening up the dark interior, making air conditioning unnecessary, and orienting the house to its wonderful natural surroundings.

With these objectives in mind, the architect, Obie G. Bowman, reoriented the entrance and opened up many areas with glass. The dramatic views are now visible through great expanses of glass—spectacular windows supported by exposed bracing and sloped glazing.

The large areas of glass bring in lots of sunlight—but heat too—and in lieu of air conditioning, ventilation chimneys were installed. This invention is used to bleed off summer heat buildups from the southwestern glazing. Such chimneys are an effective system for cooling a house with no energy cost. The technology relies on the fact that hot air rises, and the ventilation chimneys draw air up and out. (Other similar systems involve operable skylights or attic windows, combined with fans below.) The architect estimates that the cost of the ventilation chimneys and solar glazing amounted to about the same cost as any mechanical air conditioning system. But of course, the new approach is cost-free for the future.

Wood was the dominant material—both inside and out. Douglas fir was used in the interior and redwood on the exterior walls; the wood was left unfinished to age and weather naturally. There are also wood oddities like tree trunks in the interior and a "profile door," as well as extensive use of open framing, beams, and built-in shelving.

While only 850 square feet were added to the footprint of the house, the interior was entirely reworked, though the original foundation and framing were retained. There is now a split function living room with a fireplace and a new glazed octagonal conversation area. A new recreation level was added below the ground floor. The reworked house is now open and airy and has a tower-like space "that accepts sun deep into the house."

The 1971 Sonoma Coast house, with several "haphazard" remodels, and its newer version with reworked design and green solutions (opposite, bottom).

Ventilation chimneys, used to bleed off summer heat build-up, replaced air conditioning.

Newly glazed large windows are supported by Douglas fir log columns and timber girts. The newly braced frame and tower-like form brings sunlight deep into the house.

The imaginative use of wood and an emphasis on interesting spaces in the interior include a "profile" door, exposed framing, tree trunks, and multi-level design.

40
Redoing a Hurricane-Damaged Home

The Epsten Group, Inc. Architects

When this house was severely damaged by a 2004 hurricane, the architects at Epsten Group took the opportunity to make it ecologically sound, as well as to add a recreation room and update the kitchen. They won the State of Florida's first LEED Homes Silver award for the finished project.

Of major importance was how to save the massive oak tree next to the house. The answer was to build around it; the house revision is an L-shaped building with a separate 425 square foot center structure at right angles to it. The new addition includes a recreation room. (Each part of the building has its own heating and ventilating and air conditioning system.) One advantage of the design is the resulting courtyard.

The rebuilding on a slab was designed with green materials in mind at all times. Thus the wood and water-based wood finishes are urea-formaldehyde free, the floors include the use of such sustainable materials as cork or bamboo or concrete (occasionally stained to look like slate). The windows are Energy Star rated, double-paned and Argon filled, with low-E coating,, and paint used is low or No-Voc, and kitchen countertops were manufactured stone. The exterior surfaces are acrylic stucco over concrete. The hot water heater is solar (with panels mounted on the roof); there is a backup of an energy-efficient gas water heater.

This award-winning design gives full attention to the smallest details of green building—from rain chains to collect water, to retractable screening and non-adhesive tile.

A diagram of the redone house shows how it was built around the massive oak tree.

The L-shaped revision of the house includes a separate, unattached building to create a courtyard.

Large, energy-efficient new windows are used throughout.; A right-angled addition includes a separate 425 square foot recreation room with its own entrance.

The kitchen uses green materials, including manufactured stone countertops and island, low VOC wood finishes, and formaldehyde-free cabinets.

41
A Kitchen Remodel Designed for Living (and Deconstruction)

Leger Wanaselja, Architecture

Photos by Robert Kolderup and Emily Hagopian

A modest kitchen remodel in the Dog Patch section of San Francisco took a tiny 1893 "cottage" space, opened it up, and created a new, vibrant, light-filled kitchen and dining area. With only a very small space available for an addition, Leger Wanaselja Architects built a seventy-eight square foot addition into the tiny side yard, took over a space-wasting hallway, removed a dropped ceiling—and even added a (mostly recycled) redwood deck to the cottage. And there is still enough room for a vibrant permaculture garden. Tall windows and a Kal-wall skylight filled with Aerogel insulation add daylight. (The Aerogel modulates both sunlight and heat and eliminates glare.) The newly heightened ceiling allows for vertical storage.

The ecologically-sound kitchen has both recycled and eco-sensitive materials. In addition to engineered or recycled lumbers, salvaged cedar shelves, natural wood finishes, and no-Voc paints and caulk, the kitchen has another green aspect: it is designed in modular units for deconstruction without damage to the rest of the house.

In this era of expansive building, this renovation is an example of very small—but important—changes that can be made to a house that would be considered miniature by many designers. In fact, the reconstruction enabled the owners to stay in their home, with a ecologically sound and updated renovation, as well as with newly bright and inviting living spaces.

A "cottage" in San Francisco's Dog Patch neighborhood needed a new kitchen but had little room for expansion.

A seventy-eight square foot addition was placed in the tiny yard and, with the space from an unneeded hallway, created room for the new kitchen.

A skylight filled with Aerogel insulation moderates both sunlight and heat.

The charming, finished kitchen is ecologically sound with salvaged woods and low-Voc paints and finishes—and it is even designed for deconstruction.

42

Saving Energy:

A Passive and Active Solar Addition

Courtney Miller Architects

When adding to a simple 1000 square foot cape house, the aim was to increase the living space and to produce an energy-saving plan for the entire house. Architect Courtney Miller's design relied on both active and passive solar energy. The new 2000 square foot addition was designed with southern orientation and high-grade insulation.

Placed against the south wall of the existing house, the addition had passive solar gain through its south-facing windows, as well as active solar gain through roof panels. (Old growth trees, however, were saved because their "cooling effect in summer far exceeds the solar contribution that might be gained from cutting them down.")

Another major element of the thermal envelope was insulation. The entire home (both the original and new addition) was wrapped in one-and-one-half inches of rigid foam polyiso insulation. The interior walls had soy-based insulation. An air-to-air heat exchange system, as well as radiant floor heat keeps the house cozy in winter.

For summertime, air conditioning was made unnecessary by the installation of a whole house fan that sucks heat out of the house. Other details of the energy envelope include an airtight attic ladder, a Munchkin boiler (one of the smallest and most efficient of gas absorption boilers), and a solar mechanical room.

This project was shown on "Renovation Nation," a feature of the Discovery Channel, which emphasized the "passiv haus." (the original concept developed in Germany).

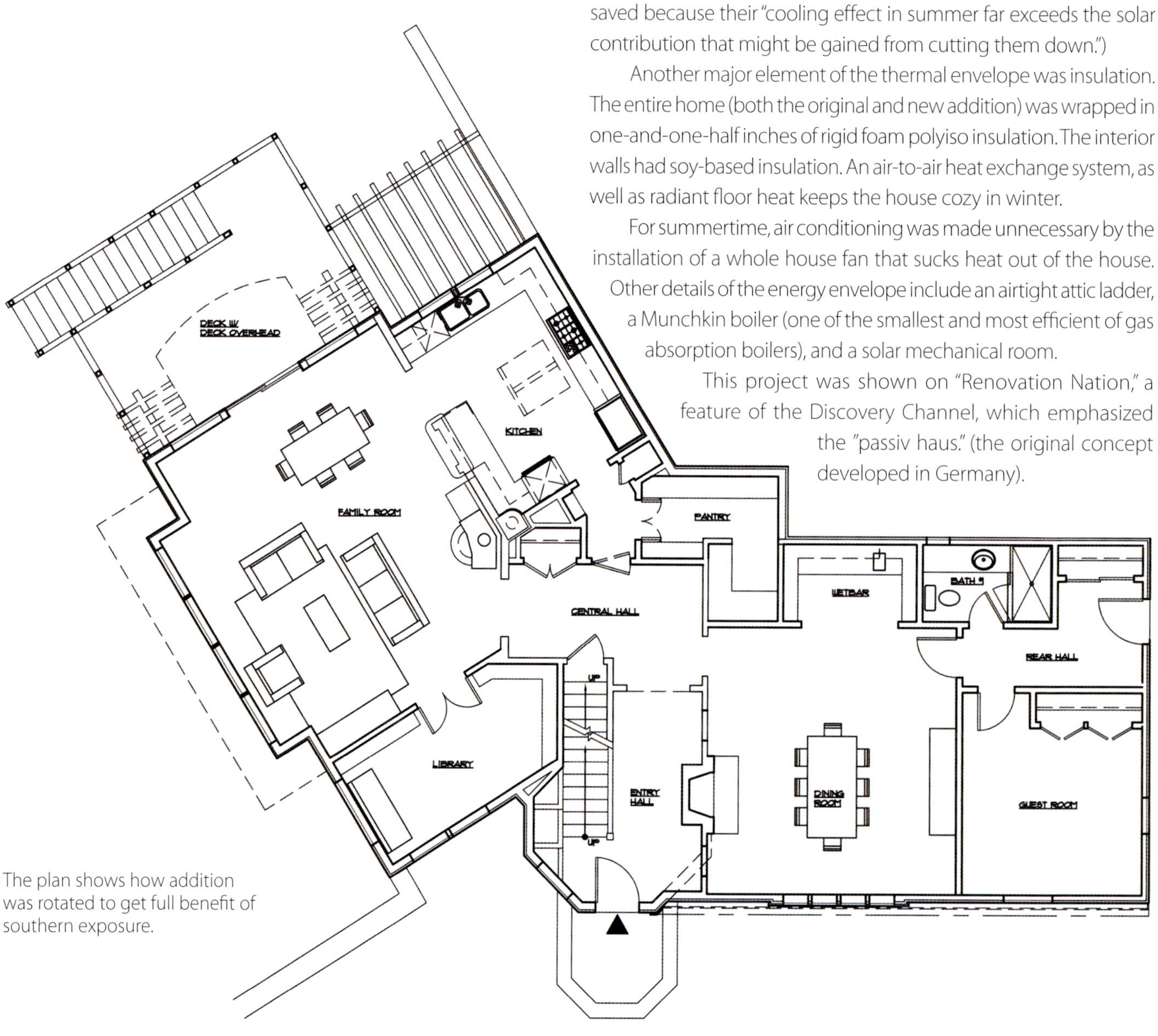

The plan shows how addition was rotated to get full benefit of southern exposure.

Solar panels on the roof produce active solar energy, while south-facing windows create passive solar gain.

Large trees around the house have "a cooling effect in summer that exceeds the solar contribution of cutting them down."

The interior is air-tight (note attic ladder in ceiling) and super-insulated; a whole house fan sucks heat out of the building making air conditioning unnecessary.

43

A Proposal for a Green Reworking of a Greek

Levenson McDavid Architects PC

A "passive house proposal" is how Levenson McDavid Architects describe their plans to rework a brick 1846 Greek Revival townhouse in Brooklyn, NY. On an historic block that is carefully monitored for any changes that might interfere with the preservation of each and every building, this house has to look exactly as it has for a century and a half.

But the proposal calls for extensive changes nonetheless: changes in insulation, air quality, thermal enclosure, and numerous additional passive attributes that will turn the 19th century house into a 21st century eco-house. "The potential synergy between our desire to preserve a building's historic heritage while making it truly energy-efficient," as the architects describe the project, shows us what can be done with meticulous planning and imagination,

The house is a four-story, two-family townhouse with a brick façade above a brownstone base. Its needs included air-tight new windows that are historically accurate in style, new insulation, repairs to the façade, new heating and cooling systems designed to reduce bills by some 90%, lower dust levels, and numerous other smaller eco-changes. (This plan will create what is known as a PHPP (Passive House Planning Package: a standard measure of the performance of green buildings.)

The first consideration is preservation: the historic façade and the decorative cornices, roof lines and window design are to be carefully updated. The thermal envelope includes non-toxic cellulose insulation in roof and walls with a vapor-open assembly (to allow moisture to dry and eliminate the potential for mold), rigid insulation for the slab, and new airtight, triple-glazed, passive solar heating windows, and continuous air barriers in the interior. Continuously filtered fresh air at low velocities throughout the house is designed for seasonal changes, while central air-cooled heat pumps will provide cooling and heating.

It should be noted that this type of renovation is not inexpensive. The basic coast of the construction is estimated at $1.2 million (with the additional requirements to meet PH standards costing approximately $80,000.) But the operational savings will be considerable; instead of a $1000 utility cost the estimate is around $300 per month.

Revival Townhouse

The front and back of the existing town house.

An architectural rendering of the renovations, front and back, showing traditional Greek Revival street side with decorative cornices and brick façade.

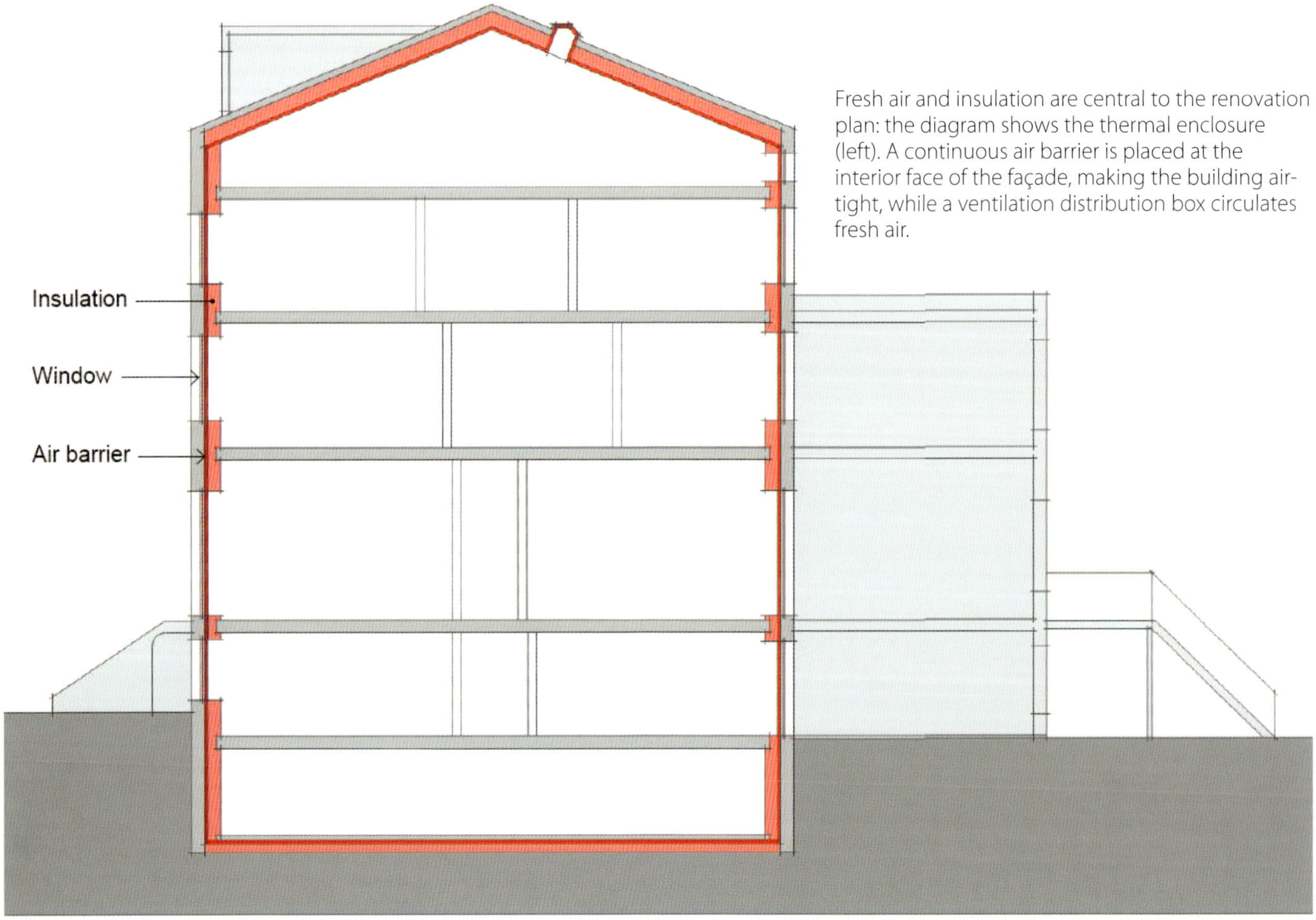

Fresh air and insulation are central to the renovation plan: the diagram shows the thermal enclosure (left). A continuous air barrier is placed at the interior face of the façade, making the building air-tight, while a ventilation distribution box circulates fresh air.

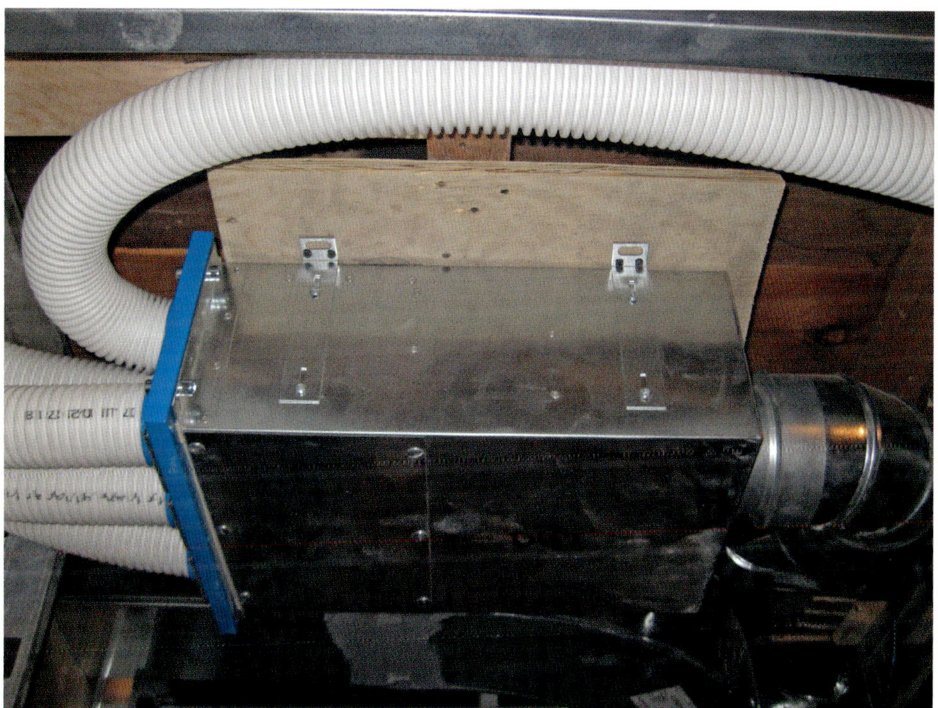

44
An Indoor/Outdoor Slab for a Passive Solar Update

Claremont Environmental Design Group, Architects

The family in this small 1920s house needed more space and wanted a remodel that would be as "earth-friendly as possible." The architects came up with a new passively solar addition and an environmentally friendly redo of the old house. Most of the original building (including windows and doors removed during the renovation) was salvaged or reused. The addition included a new kitchen, bedrooms, a redone patio—and the centerpiece—a new passive solar family room.

The design of the family room involved the ingenious use of a "slab on grade," which functions as a thermal mass that includes both indoor and outdoor areas. The slab is heated by the sun through a south wall of patio doors (which bi-fold open) and—with the adjoining patio space—can almost double the size of the room. The patio's 4 x 12 foot trellis beams extend from the patio across the family room, while the sides of the patio are enclosed on either side by new rooms. The east and west patio walls protect the family room from overheating (and define the patio space). A pond is integrated into the patio's southern edge; it helps condition the air before it enters the house.

The house also has more daylight in the interior with the use of the patio doors, a new skylight, and a high eyebrow window that "bounces light off the north kitchen wall." This house is thus able to use the benign California climate and bright light with a passive solar system that encourages outdoor activity and cuts down on indoor energy costs.

The original 1920s California house has a garden setting.

The new patio's 4 x 12 foot trellis beams extend into and across the family room.

The slab floor is used as a thermal mass which is heated by the sun through the south wall's bi-fold patio doors.

A small pond (with lily pads) at the edge of the patio helps to condition the air as it enters the home.

45
An Old House Inside a New One

Webber + Studio, Inc. Architects

Photos by Paul Bardagiy

Beginning with a modest one-story frame house in Texas, this renovation and addition totally reworked both space and design. In order to accommodate an expanding family and to add a father-in-law suite, the plan reworked the original house with a variety of new spaces by surrounding it and building around the perimeters of the original house.

New outdoor areas included a small interior court, a large backyard, a terrace, and as an effective cooling design for a warm climate—a kind of dogtrot. (A dogtrot is a roofed but open-ended passageway from one outdoor area to another.) The architects were able to create what they describe as an "interlocking relationship between the house's masses and resultant spatial voids." The design blocks the strong western sun and provides more desirable northern and southern exposures.

The manipulation of the skin" of the redone building created a modern-looking and well-insulated exterior that can withstand the harshest exposure. The building was wrapped in a new exterior; materials "with a strong regional presence" include "regionally ubiquitous metal," and locally prevalent cypress on some exterior walls. Pecan wood and limestone are used in the interiors. There is also a polycarbonate wall.

The plan created a number of new spaces, including a car-park and bay projections, as well as a soaring 25-foot tall space with an adjacent 8-foot tall sitting area. In salvaging the original house for imaginative reuse, as well as ecologically sound planning and construction with local materials, this project meets many of the criteria for green renovation.

The "before" picture shows the original building: a modest one-story frame house.

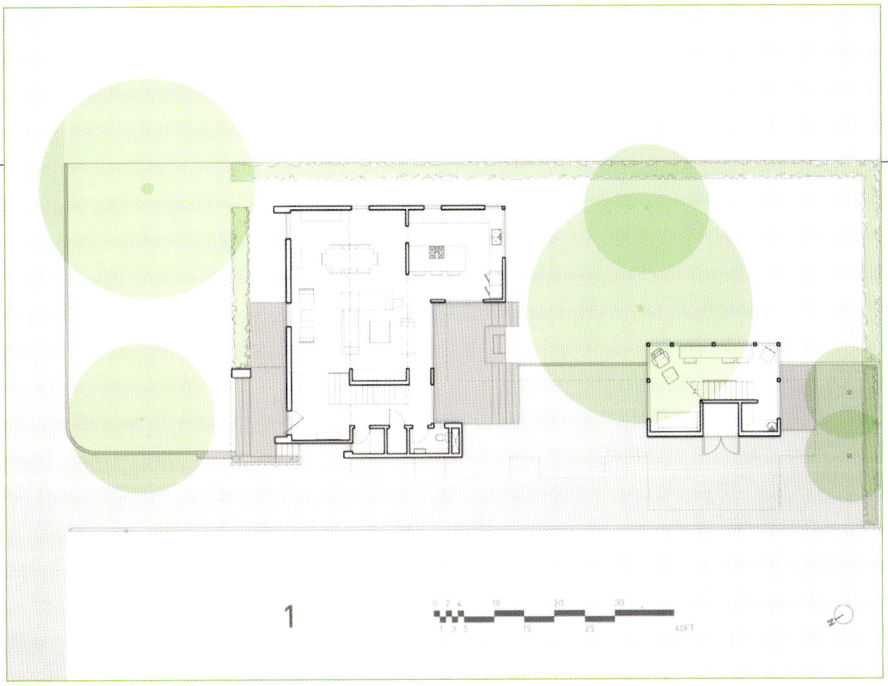

The floor plan shows how the existing house was retrofitted with new construction built around the original building.

The redone two-story house has an exterior that includes metal cladding and cypress wood.

The architects created various outdoor spaces, including a small court, a spacious backyard, a terrace, and a dog-trot inspired breezeway.

The interior uses local prevalent pecan wood and limestone, as well as polycarbonate walls.

The Architects and Designers

Arkin Tilt Architects
1101 8th Street Suite 180
Berkeley, CA 94710
Tel: 510-528-9830
Web: www.arkintilt.com

Bercy Chen Studio LP
1111 East 11th Street Suite 200
Austin, TX 78702
Tel: 512-481-0092
Web: www.bcarc.com

Obie G. Bowman FAIA
P.O. Box 1114
Healdsburg, CA 95448
Tel: 707-433-7833
Web: www.obiebowman.com

Centerbrook Architects and Planners
Charles W. Moore and Mark Simon
67 Main Street
P.O. Box 955
Centerbrook, CT 06409
Tel: 860-767-0175
Web: www.centerbrook.com

Claremont Environmental Design Group (CEDG)
480 North Indian Hill Blvd.
Claremont, CA 93454
Tel: 909-625-3916
Web: www.cedg-design.com

Taylor Dawson and Associates, LLC Architecture
310-21st Street North Suite 100A
Birmingham, AL 35203-3364
Tel: 205-328-4455
Web: www.taylordawson.com

Angela Dean, AIA
AMD Architecture
311 South 900 East
Salt Lake City, UT 84102-2310
Tel: 801-322-3053
Web: www.amdarchitecture.com

EDGE Architects (Environmental Design Group Enterprise)
Ken Anderson and Pamela Freund
Box 2489
Taos, NM 87571
Tel: 575-758-5642
Web: www.mail@edgearchitects.com

Ryan Enschede Studio
375 Lincoln Place 4C
Brooklyn, NY 11238
Tel: 347-663-6898
Web: www.ryanenschede.com

The Epsten Group Inc.
429 Edgewood Avenue
Atlanta, GA 30312
Tel: 404-577-0370
Web: www.theepstengroup.com

Christina Griffin Architect PC AIA
10 Spring Street
Hastings-on-Hudson, NY 10706
Tel: 914-478-0799
Web: www.christinagriffinarchitect.com

David Hertz FAIA Architects Inc.
Studio of Environmental Architecture
1920 Olympic Blvd.
Santa Monica, CA 90404
Tel: 310-829-9932
Web: www.STUDIOEA.com

Robert Knight Associates Architects
157 Hinckley Ridge Road
P.O. Box 803
Blue Hill, ME 04614
Tel: 207-374-2845
Web: www.knightarchitect.com

Lake/Flato Architects
311 Third Street
San Antonio, TX 78205
Tel: 210-227-3335
Web: www.LAKEFLATO.com

Leger Wanaselja Architecture
2320 McGee Avenue
Berkeley, CA 94703
Tel: 510-848-8901
Web: www.greendwellings.com

Levenson McDavid Architects PC
45 Park Place
Brooklyn, NY 11217
Tel: 718-622-1600
Web: www.imarch.com

Locus Architecture
708 West 40th Street
Minneapolis, MN 55409
Tel: 612-701-7675
Web: www.locusarchitecture.com

Courtney Miller Architects
2401 Ivy Drive Suite 100
Oakland, CA 94606
Tel: 781-646-6165
Web: www.ecobuild.com

Ted Montgomery
Groundswell Architects PC
477 Ten Stones Circle
Charlotte, VT 05445
Tel: 802-425-7717
Web: www.groundswellarchitects.com

Mark O'Bryan, Architect
The Design Studio
900 Distillery Commons
Louisville, KY 40206
Tel: 502-566-6006
Email: m2@iglou.com

Pugh and Scarpa Architects
2525 Michigan Avenue
Santa Monica, CA 90404
Tel: 310-828-0226
Web: www.pugh-scarpa.com

Richard Renner/Architects
35 Pleasant Street
Portland, ME 04104
Tel: 207-773-9699
Web: www.rrennerarchitects.com

John Ringel
Jersey Devil
Box 280
Stockton, NJ 08559
Tel: 609-397-1806
Web: www.johnringel@earthlink.net

Rick Harlan Schneider AIA
ISTUDIO Architecture
1611 Connecticut Avenue, 200A
Washington, DC 20009
Tel: 202-758-2037
Web: www.lstudioarchitecture.com

Shelter Architects
2913 Harriet Avenue Suite 102
Minneapolis. MN 55408
Tel: 612-870-4081
Web: www.shelterarchitecture.com

Jennifer Siegal
Office of Mobile Design
1725 Abbot Kinney Blvd.
Venice, CA 90291
Tel: 310-439-1129
Web: www.designmobile.com

Ann Temple, Architect
Schroeder Murchie Niemiec Gazda-Auskalnis Architects, Ltd.
936 West Huron Street
Chicago, IL 60642
Tel: 312-829-3355
Web: www.smng-arch.com

Andrew Tesoro
Tesoro Architects
1841 Broadway Suite 1200
New York, NY 10023
Tel: 212-757-1841
Web: www.TESOROARCHITECTS@AOL.com

Elizabeth Thompson, Architect
250 Mercer Street
New York, NY 10012
Tel: 212-614-7180
Web: www.elizabeththompsonarchitect.com

Tigerman McCurry Architects, Ltd.
444 North Wells Street #206
Chicago, IL 60654
Tel: 312-644-5880
Web: www.tigerman-mccurry.com

Webber + Studio
360 West Avenue Suite 1322
Austin, TX 78701
Tel: 512-236-1032
Web: www.webberstudio.com

Peter Kurt Woerner and Associates
Studio 44
44 Kendall Street
New Haven, CT o6512-1707
Tel: 203-466-1923
Email: pwoerner@sbcglobal.net

Organizations and Programs To Contact

Alliance to Save Energy: **www.ase.org**

American Solar Energy Society: **www.ases.org**. A guide to resources regarding renewable energy.

Architects, Designers and Planners for Social Responsibility: **www.adpst.org**. A way to choose people who have eco-renovation at the forefront of their planning.

Building Materials Reuse Association: **www.BMRA.org**. Listed by state.

Build It Green: **www.build-green.org/guide**. A guide to materials that are ecologically sound.

Department of Energy: **www.eere.energy.gov/greenpower**. Contains material on a variety of programs relating to energy saving.

Earth Pledge: **www.earthpledge.org**. An organization dedicated to green roofs and sustainability initiatives.

Ecological Building Network: **www.ecobuildnetwork.com**

Energy Star: **www.energystar.gov.** A federal program in which energy use is calibrated.

Green Home NYC: **www.greenhomeNYC.org.** A non-profit group that advises on ecological methods of building.

LEED (Leadership in Energy and Environmental Design): **www.leed.org**. A rating system of the United States Green Building Council, it rates buildings on how much energy they use or waste.

National Association of Homebuilders: **www.nahb.com**. They have a certification system for green building.

The Rutgers Center for Green Building: **www.greenbuilding-rutgers.us**. This organization has links to national resources, including rebates for solar and wind projects.

United States Forest Stewardship Council: **www.fscus.org**. The seal of this organization shows that wood is harvested properly, using sustainable methods.